HUMAN NATURE AND THE LIMITS OF SCIENCE

JOHN DUPRÉ warns that our understanding of human nature is being distorted by two faulty and harmful forms of pseudo-scientific thinking. Not just in the academic world but increasingly in everyday life, we find one set of experts seeking to explain the ends at which humans aim in terms of evolutionary theory, and another set of experts using economic models to give rules of how we act to achieve those ends. Dupré charges this unholy alliance of evolutionary psychologists and rational-choice theorists with scientific imperialism: they use methods and ideas developed for one domain of inquiry in others where they are inappropriate. He demonstrates that these theorists' explanations do not work, and furthermore that if taken seriously their theories tend to have dangerous social and political consequences. For these reasons, it is important to resist scientism—an exaggerated conception of what science can be expected to do for us. To say this is in no way to be against science—just against bad science.

Dupré restores sanity to the study of human nature by pointing the way to a proper understanding of humans in the societies that are our natural and necessary environments. He shows how our distinctively human capacities are shaped by the social contexts in which we are embedded. And he concludes with a bold challenge to one of the intellectual touchstones of modern science: the idea of the universe as causally complete and deterministic. In an impressive rehabilitation of the idea of free human agency, he argues that far from being helpless cogs in a mechanistic universe, humans are rare concentrations of causal power in a largely indeterministic world.

Human Nature and the Limits of Science is a provocative, witty, and persuasive corrective to scientism. In its place, Dupré commends a pluralistic approach to science, as the appropriate way to investigate a universe that is not unified in form. Anyone interested in science and human nature will enjoy this book, unless they are its targets.

Human Nature
and the
Limits of Science

JOHN DUPRÉ

CLARENDON PRESS · OXFORD

*This book has been printed digitally and produced in a standard specification
in order to ensure its continuing availability*

OXFORD
UNIVERSITY PRESS

Great Clarendon Street, Oxford OX2 6DP

Oxford University Press is a department of the University of Oxford.
It furthers the University's objective of excellence in research, scholarship,
and education by publishing worldwide in

Oxford New York

Auckland Cape Town Dar es Salaam Hong Kong Karachi
Kuala Lumpur Madrid Melbourne Mexico City Nairobi
New Delhi Shanghai Taipei Toronto
With offices in
Argentina Austria Brazil Chile Czech Republic France Greece
Guatemala Hungary Italy Japan South Korea Poland Portugal
Singapore Switzerland Thailand Turkey Ukraine Vietnam

Oxford is a registered trade mark of Oxford University Press
in the UK and in certain other countries

Published in the United States
by Oxford University Press Inc., New York

ISBN 0-19-926550-X

For Emily Farahar (1979–2001)
whom science failed

Acknowledgements

In her recent book *The Dappled World* (1999), Nancy Cartwright makes the first reference in print of which I am aware to the Stanford School of the philosophy of science. The fixed point around which this movement revolved was undoubtedly Patrick Suppes. Suppes pioneered many of the themes we took up and taught at Stanford for several decades. Meanwhile, Ian Hacking, Peter Galison, Cartwright, and myself, taught there for various periods. Suppes's recent retirement completes the diaspora. However, these remain among the contemporary philosophers of science with whom my work has the greatest affinities, and I am grateful to Stanford University for bringing us together.

This book was written over the best part of a decade, and I cannot hope to recall all the colleagues and students with whom, at various times, I have discussed some of the topics it addresses. Colleagues at Stanford, Birkbeck, and Exeter have provided encouragement, advice, and intellectually stimulating environments that have made my work better. Ancestors of several chapters have been presented to audiences in those and many other universities in the United States and Europe, and I have benefited from the comments of many people on those occasions. Adrian Haddock, Jonathan Kaplan, Philip Kitcher, Helen Longino, Ina Roy, and an anonymous referee each read an entire draft of the book, and the final version has been much improved by their thoughtful comments. My understanding of biology owes much more to the work of Ernst Mayr and Richard Lewontin than appears from the citations, which focus inevitably on those with whom I disagree. Peter Momtchiloff of Oxford University Press has been an enthusiastic and supportive editor and has enabled publication to proceed with an efficiency and dispatch not universal in academic publishing.

The book was completed during a period of leave funded by the Arts and Humanities Research Board and research leaves granted by

Birkbeck College, University of London and the University of Exeter. I would like to thank all of these institutions for their support.

From my sons, Gabriel and Julian, I learned much of what I know about human development and about the futility of the dichotomy between nature and nurture. And, as always, my greatest debt is to Regenia Gagnier. Her humanistic learning, critical virtuosity, and domestic best practice have added value to my environment for almost twenty years.

I have drawn freely on previously published work in writing this book, and I would like to thank the publishers of various papers for allowing me to include substantial extracts in this book. Parts of Chapters 2, 3, and 4 are derived from 'Normal People', *Social Research* 65 (1998): 221–48; 'Against Reductive Theories of Human Behaviour', *Proceedings of the Aristotelian Society*, supp. vol. 72 (1998): 153–71, reprinted by courtesy of the editor of the Aristotelian Society, copyright © 1998; and 'What the Theory of Evolution Can't Tell Us', *Critical Quarterly* 42 (2000): 18–34. Chapter 5, part 8 is based on 'The Fight for Science and Reason', a review of Norman Levitt's *Prometheus Bedevilled*, *The Sciences*, March/April 2000: 40–5. Chapter 6 draws on 'On Scientific Imperialism', *Philosophy of Science Association Proceedings* 2 (1994): 374–81; copyright © 1994 by the Philosophy of Science Association. All rights reserved. Chapter 6, part 6 uses material from two papers published jointly with Regenia Gagnier: 'On Work and Idleness', *Feminist Economics* 1 (1995): 1–14, copyright © IAFFE 1995, and 'A Brief History of Work', reprinted from *Journal of Economic Issues* 2 (1996): 1–14, by special permission of the copyright holder, the Association for Evolutionary Economics. Chapter 7 includes most of 'The Solution to the Problem of the Freedom of the Will', *Philosophical Perspectives* 10 (1996): 385–402.

Contents

1. Introduction   1

2. The Foundations of Evolutionary Psychology   19
   1. Introduction   19
   2. Overview of the Evolutionary Argument   21
   3. How Much Must Evolution Explain?   23
   4. Atavism   25
   5. Do Brains Cause Behaviour?   31
   6. Nature and Culture   38

3. The Evolutionary Psychology of Sex and Gender   44
   1. Introduction   44
   2. The Sociobiology of Sex and Gender: The Classic Story   45
   3. Sociobiology Twenty-five Years Later   48
   4. Further Reflections on the Poverty of Evolutionary
      Psychological Inference   62

4. The Charms and Consequences of Evolutionary
   Psychology   70
   1. Introduction   70
   2. The Epistemological Charms of Evolutionary Psychology   70
   3. The Sociological Appeal of Evolutionary Psychology   81
   4. Political and Ethical Implications   85

5. Kinds of People   93
   1. Introduction   93
   2. The Power of Culture   95
   3. Cultural Change: Anagenetic Evolution   98
   4. Cultural Species: Cladogenetic Evolution   99
   5. Humans and Other Species   102
   6. Cultural Species Again   107
   7. The Value and Future of Cultural Diversity   109
   8. Imperialist Scientism   113

6. Rational Choice Theory   117

    1. Introduction   117
    2. *Homo Economicus*   120
    3. Problems with Imperialist Science   128
    4. Central Themes in Scientistic Methodology   132
    5. A Comparison of Imperialism in Evolutionary Biology
       and Neoclassical Economics   137
    6. Simplistic Economics vs. Sophisticated Pluralism:
       The Case of Work   138
    7. The Normative and the Positive   146

7. Freedom of the Will   154

    1. Introduction   154
    2. Free Will and Determinism   155
    3. Microphysical Determinism and the Causal Inefficacy
       of Everything Else   159
    4. Causal Incompleteness   163
    5. Machines and Organisms   170
    6. Moral Autonomy   177
    7. Conclusion   182

Bibliography   189

Index   197

1
Introduction

While working on this book I happened to turn on the third instalment of a television series on human hormones. The official topic of this episode was Love. In between images of chemical clouds bubbling out of glands and diffusing through the body, the programme traced the effects of hormones on sexual differentiation *in utero* and in puberty. Distinguished scientists reported the exciting and sometimes surprising results of our recent ability to measure the levels of hormones in bodies, and correlations between these levels and the emotional states of the subjects were noted. As different behavioural tendencies were shown to develop in males and females, evolutionists informed us about the functions these might have served for our Stone Age ancestors. Reaching the official topic of love, we were taught to distinguish its various phases—infatuation, obsession, companionship—and their hormonal correlates. Magnetic Resonance Imaging of obsessed lovers revealed similarities between their brain activities and those of the mentally disturbed, providing, apparently, scientific evidence that love is indeed a form of madness. Later, we learned that whether male voles remained faithful to their partners or indulged in untrammelled promiscuity depended on the presence of specific hormones, and we were invited to speculate as to whether similar mechanisms might operate in humans. And so on. Although much of this work was admitted to be at a somewhat speculative stage, the scientists involved expressed no reservations about the possibility that love might turn out to be caused by, or just to be—such ontological subtleties were not addressed—a sequence of hormonal surges; nor did members of the public asked to comment on some of these scientific claims, though some expressed the view that the topic of love should be left to poets, and that these scientific facts were better left unknown.

This programme illustrates the hold on our culture of what I call scientism, an exaggerated and often distorted conception of what science can be expected to do or explain for us. One aspect of scientism is the idea that any question that can be answered at all can best be answered by science. This, in turn, is very often combined with a quite narrow conception of what it is for an answer, or a method of investigation, to be scientific. Specifically, it is supposed that canonical science must work by disclosing the physical or chemical mechanisms that generate phenomena. Together these ideas imply a narrow and homogeneous set of answers to the most diverse imaginable set of questions. Everywhere this implies a restriction of the powers of the human mind; but nowhere is this restriction more disastrous than in the mind's attempts to answer questions about itself. This topic, the effect of scientism of this sort on attempts to understand the human mind and the human behaviour through which the mind is displayed, is what this book is about. Unsurprisingly, perhaps, misguided approaches to the understanding of human behaviour, or human nature, are fraught with danger.

It is impossible to attend to the contemporary mass media without hearing of the genes for this or that feature of human physiology or behaviour that scientists have discovered or are on the verge of discovering. Perhaps most widely publicized are the genes predisposing us to cancer or heart disease, discoveries which may indeed one day offer help in enabling us to deal more effectively with the major killers of contemporary Western people. But more exciting journalism is offered by the genes for aggression, alcoholism, homosexuality, promiscuity, rape, intelligence, criminality, and so on, genes that purport to explain the great variety of human behaviour.[1] And scientists and journalists seem more than willing to collaborate in the production of such provocative news. Slightly less obvious, but easily available to the discerning broadsheet reader, are the reports of the scientists who claim to know why such genes exist. These are the sociobiologists, now preferring to be known as evolutionary psychologists, who entertain us with stories of how these traits, or the psychological dispositions that underlie these traits, served the reproductive interests of our Stone Age ancestors.

Another scientific perspective with which the educated Westerner is almost inevitably familiar is that of the dismal science, economics. Although there are important dissenting tendencies within the

---

[1] The scientific difficulties raised by the tendency to talk about genes for such complex human behaviours are very clearly described by Kaplan (2000).

discipline of economics, there is also an overwhelmingly dominant hegemony. This is the conception of economics as the investigation of the consequences of individuals striving to maximize their selfish interests. And recently this selfish model of human life has increasingly been imported from its homeland in commodities markets and inflation rates, and offered as a path of insight into human life generally. As important as what a scientific programme emphasizes is what it leaves out. Economics describes people's interests in terms of their particular tastes, but it does not on the whole concern itself much with where tastes come from. Tastes are typically treated as 'endogenous'. This term is crucially ambiguous. Sometimes it means that tastes are treated as given within the context of a particular model, a relatively harmless methodological convenience. But it is very easy to slide from this to the idea that tastes are endogenous in the sense that they are somehow generated by intrinsic features of the person who exhibits them. This is a remarkable view. One might suppose that people's tastes would be formed to a great extent by their culture, their family background, advertising, and so on. But this common-sense idea would be fatal to a large part of the project of contemporary economics. For economists, despite generally trying to distance themselves from normative matters, do tend to take it as obvious that satisfying people's desires is a good thing (hence the absurd tendency to describe advertising as merely the communication of information). This makes it seem worthwhile to do economics without politics, without consideration of the processes that help to determine what people want, and yet to offer the wisdom of economics to politicians, as a path to providing something good, the maximum satisfaction of people's wants.

It would be nice for economists to have access to an independent theory that helped to explain why people have the particular endogenously generated tastes they do, and here there is a natural alliance between economics and the parts of biology just mentioned. Tastes—for alcohol, sexual partners, weapons, education, and much else—can be seen as produced by particular genes, and genes that owe their existence in modern human populations to the advantages that they conferred on our distant ancestors. So these aspects of biology and of economics are well suited to one another.[2] Biologists will

---

[2] Exactly this alliance has been proposed by prominent evolutionary psychologists John Tooby and Leda Cosmides (1994).

tell us what people most fundamentally want, and economists will tell us how they will act in their attempts to get as much as possible of it. Here we have a sketch of a far-reaching scientific account of human behaviour.

The bits of science I have alluded to are, in my view, seriously misguided. Worse, they provide justification or encouragement for social policies and personal behaviour that contribute to the provision of poor environments for the flourishing of many people. The elaboration of these claims is the central task of the present work. Apart from attempting to explain in some detail what I take to be the deficiencies of these projects, I also want to show how they derive a great deal of their plausibility from a widespread, perhaps even orthodox, philosophy of science. I have devoted an earlier book (Dupré, 1993a) to the attempt to undermine the conception of science articulated by this philosophy. In this book I want to reinforce the earlier argument by exhibiting the kind of bad science to which it naturally leads. Bad science, when directed at human nature or society, is always liable to lead to bad practice. And if there is one overriding reason for people to care about dubious science, it is because it lends support to pernicious social policy.

Science, it is often said, is the religion of our era. Where once we expected priests to give us insight into the nature of the cosmos and of human existence, now we look rather to men, and sometimes women, in white lab coats. Where once public expenditure in the service of deeper truth might have taken the form of mighty cathedrals, today it will be found in cyclotrons and gene-sequencers.[3] While it is no part of the thesis of this book that we should return to this earlier age of theocratic epistemology, I shall argue that science as it has traditionally been conceived has serious limitations in its ability to answer some of the most profound questions we are given to ask and, more specifically, to answer questions about the nature and causes of human behaviour. My more positive thesis is that the only hope for serious illumination of such questions is a pluralistic one, an approach that draws both on the empirical knowledge derivable from the (various) sciences, and on the wisdom and insight into human nature that can be derived from more humanistic studies.

---

[3] The parallels between religion in the medieval period and science today are brilliantly explored by Feyerabend (1978).

The perspective underlying this positive thesis is not, however, the traditional dualistic one that sees a certain kind of scientific approach as adequate to understanding everything except the human mind. That perspective, emanating, notoriously, from Descartes, has dominated anglophone philosophy for over three centuries. Currently philosophy is dominated not by the Cartesian thesis itself, but by one of the traditional reactions to it, monistic materialism. Nowadays philosophers tend to call themselves 'physicalists' rather than 'materialists'. One reason for this terminological innovation is that philosophers wish to distance themselves from the specific conception of the material world assumed by earlier philosophers, materialist or otherwise. No one now thinks of the world, as early modern philosophers tended to, as composed of tiny billiard balls. The idea, rather, has been to tie the materialist doctrine to whatever account of the physical world physicists finally agree upon (Smart, 1978). Physicalists then hold that this final physical story about the world's ultimate constituents is, in some sense, the whole truth about the world. This idea is sometimes referred to as the thesis of the 'completeness of physics'.

I do not deny that the physics of elementary particles may very well eventually provide us with the whole truth about something, namely the nature of the stuff of which the world is ultimately composed. I do not believe that there are, in addition to the things that physicists theorize about, immaterial minds or deities. I believe, rather, that there are countless other kinds of things: atoms, molecules, bacteria, elephants, people and their minds, and even populations of elephants, bridge clubs, trades unions, and cultures. I agree with the physicalists that to the extent that these things are composed of anything they are, ultimately, composed of the entities of which physicists speak. Where I differ is in my assessment of the consequences of this minimal compositional physicalism. The truth about physical stuff, in my view, is very far from being the truth about everything.

Physicalists generally suppose that the truth of the doctrine of exhaustive physical composition has profound implications for our knowledge of the world, especially to the extent that that knowledge aspires to be scientific. Not long ago, this opinion was held by many philosophers in a quite uncompromising form. These philosophers held that the only ultimate truth about the world was the truth to be gained from physics; and that this was in a certain sense the complete truth. Against the prima facie appearance that the vast majority of knowledge concerned much larger, more complex things than were the preserve of physicists, they argued that scientific knowledge about large, complex things could be shown to be derivable from knowledge about the simple things described by physicists. This was classical physicalist reductionism. The project was seen as hierarchical. Chemistry would be derived from physics, molecular biology from chemistry, and so on up the chain of physical complexity. More recently it has been increasingly widely recognized that in practice no such reductive programme was remotely feasible. There was no way, for instance, to derive biological facts about organisms exhaustively from facts about molecules, and it has become more and more clear that there is little chance even of deriving much of chemistry from physics.

I won't discuss the details of these problems here, but rather would like to look at some of the responses they have provoked. The difficulty, simply put, was that while physics was still held in principle to embody the complete truth about the world, most of that truth was seen to be inaccessible. Much of what we already know turns out not to be translatable into appropriate truths of physics. The most ambitious response is to conclude, 'So much the worse for what we already (took ourselves to) know'. This is the position known as eliminativism, which holds that to the extent that higher-level sciences cannot be reduced to physics, or at least to a lower-level science in the structural hierarchy, they should be rejected, and ultimately replaced with scientific knowledge from the lower level. This is best known in the doctrine about the mental most conspicuously defended by Paul and Patricia Churchland (P. M. Churchland, 1995; P. S. Churchland, 1986). They argue that since much of what we currently believe about the mental looks impossible to correlate with things we know about the brain, presumably the lower-level structure responsible for mental phenomena, we should anticipate the replacement of talk about the mental with talk about the neural. Eventually we will learn to stop saying things like 'I have a pain in my knee' or 'That tastes sweet', and speak instead of the activities of appropriate groups of neurons. Since this strikes me as pure science fiction, and not the kind that has much chance of turning into fact, scientific or otherwise, I shall not discuss it here.[4]

---

[4] I do present some objections to such views in Dupré, 1993a: 148–59.

A more plausible response stops just short of replacement. It agrees that what is really going on in some ultimate metaphysical sense is purely physical, but, not anticipating either replacement or reduction, concludes that the things we say about the mental (or, for that matter, the chemical or the biological in so far as these also resist reduction) should not be taken as seriously as we might have been inclined to take them. We should, in fact, treat psychology, or biology, instrumentally rather than realistically. The claims those disciplines make, that is to say, should not be taken as literally true, but rather as constituting useful devices for interacting effectively with the biological or psychological worlds. It does seem to me correct to argue that, if indeed physics is complete—in other words, if there is a set of physical laws and a set of facts about the properties of physical particles such that when these are fixed everything is fixed—then this instrumentalist picture is the strongest endorsement we can offer for the non-physical sciences. Either the claims of biology, for instance, follow from this set of physical truths or they do not. If they do not, then they cannot be part of the whole truth, which is to say they cannot be true. If they do, then they are, in principle at least, reducible; all that prevents us from carrying out the appropriate reductions is the limited cognitive or computational capacities of our brains.[5]

Although, as I have said, strong forms of reductionism have proved unworkable as a practical way of doing science, the underlying picture that reductionism expresses continues to exercise a profound influence on science. It is still common, for instance, to conceive of genuinely scientific explanation as being necessarily mechanistic. That is to say, the task of science is seen as one of showing why things behave as they do by disclosing the way their constituent parts interact to produce that gross behaviour. Even if we are unable to go right down to the finest level of structure, the microphysical, we can move in that direction by such mechanistic explanation. I do not deny that there is an important role for such mechanistic explanations, and that some of the greatest of scientific achievements are of this character. Mechanism, I take it, has proved wonderfully successful at addressing questions about how things work. But when it is taken beyond this limited, if important, role and inflated into a general metaphysical world view, it is disastrous. For

[5] A good example of such a view is the treatment of biology by Rosenberg (1994).

we do not only want to know how things work, we want to understand what they do, and why. And such questions can usually only be answered by looking at the context in which a thing is situated, and the interactions it is engaged in with other things. The distortions of the mechanistic view acquire a further dimension when applied to the behaviour of living things. For one of the things that living things do is grow, or develop. And this, like other things they do, is a process dependent on countless and complex interactions with their environments. As I shall explain in the following chapters, the wholly confused idea that the development of an organism is merely the implementation of a plan or the running of a program somehow written in the DNA, is a paradigm of the consequences of mechanistic distortion.

Evolutionary psychology extends this error to the very most complex subject matter, the human brain. It has only to be stated to be obvious that the human brain develops in partial response to vast numbers of environmental influences, and though evolutionary psychologists are not so foolish as to deny this, they go to great lengths to minimize its importance. One of my positive aims in this book, in contrast, is to emphasize not only the importance of these interactions, but also their remarkable character. I shall claim that humans are, in important respects, constituted by the social context in which they exist: the capacities that they possess depend not just causally but constitutively on facts about their social contexts. And in the final chapter of the book I argue that it is in the relationship between the individual and society that we find the basis of what genuinely deserves to be called individual human autonomy.

These positive theses will be developed at various points in the book, but the main focus will be critical, an exploration of the deep deficiencies of scientific projects motivated by the monistic and reductive metaphysics I reject, as well as of the ways in which this metaphysics works to legitimate these misguided and unsuccessful scientific projects.

I don't propose in the present work to address systematically the deficiencies of the mechanistic, physicalist metaphysics. I have already undertaken this task at book length (Dupré, 1993a), and my hope here is rather that it shall be known by its fruits. It may, however, be helpful at the outset to summarize some of the objections I have to this very widespread philosophical viewpoint, and, in particular, to say a little about why I reject the thesis of the completeness of physics.[6] One central point is simply that science is a credible source of knowledge to the extent that it is empirical; and there is not a shred of evidence for the completeness of physics. One might indeed wonder whether there possibly could be such evidence. But whereas it would certainly be impossible to provide conclusive evidence for the completeness of physics (an instance of a familiar point about negative existence claims), it would surely be possible to find partial evidence. And even this, I suggest is lacking. What I have in mind as partial evidence would be evidence that physics was able to explain something significantly removed from its own original subject matter. There is, in fact, some debate about whether quantum mechanics can explain significant parts of chemistry. It is sometimes said that properties of simple molecules can be so explained, and sometimes denied. But no one claims to have explained the properties of complex molecules—DNA or polypeptide molecules, say—from quantum mechanics. The question, then, seems in the end to be only one of the sharpness of the boundaries of physics. Perhaps the methods and theories of physics are useful some way into what is traditionally thought of as chemistry, perhaps not. But certainly they are of no use where chemistry begins to merge into molecular biology. It will characteristically be objected at this point that the impossibility of reductive explanation to which I here refer is 'merely practical'. Our minds, once again, lack the necessary computational capacities. But I won't dwell on this response, because the argument here is about the empirical status of claims for the scope of physics. And whatever the status of claims for truths that exceed our computational capacities, they are surely not empirical (for us, anyhow, and who else is there?).

Another common move is to suggest that reductionism is misconceived as a practical goal, but is vital as a regulative ideal of science.[7] In accordance with this regulation, science should, on such a view, aim to explain phenomena as far as possible in terms of the properties and behaviour of lower-level constituents. This

---

[6] Some more detailed arguments for this point are provided in the final chapter of this book.

[7] The point might be developed further by suggesting that science thus presupposes the completeness of physics, and since the success of science is evidence for the truth of its presuppositions (inference to the best explanation, perhaps), the success of science is evidence for the completeness of physics. However, I not only reject the premise of this argument, but am also sceptical about its form.

methodological norm can remain neutral, and perhaps even a bit fuzzy, about its relation to the sorts of metaphysical theses just mentioned. But unlike them, it can do real work in guiding science towards certain projects rather than others. This methodological reductionism will favour explanation of phenomena in terms of their intrinsic properties over explanations that emphasize the contextual or environmental influences on them. I say, rather vaguely, 'emphasize' because it is really clear that any explanation must appeal to both internal and contextual features. Intrinsic properties of things will affect their behaviour differently in different contexts: it is perhaps an internal property of a glass that it is fragile, but this only explains, or predicts, its breakage when it is struck or dropped.

Although this may seem, therefore, a rather arbitrary distinction, it is in fact of great importance. There is a sense in which much science really does emphasize the intrinsic over the contextual. Many scientific explanations work by providing models that treat only selected features of a situation, and assuming that nothing else significant interferes (the so-called *ceteris paribus* condition). Since intrinsic features are generally more stable than contextual ones, there are obvious advantages for models that select those features. There are paradigm cases in which this is a highly effective strategy. Perhaps the best possible example is the one generally considered responsible for the take-off of modern science, the Newtonian model of the solar system. The masses, positions, and velocities of a small number of objects and the law of gravitation are essentially the only factors that matter in determining the trajectory of this system. Although, as is well known, even this system is currently computationally intractable, approximate methods for analysing the system are good enough to give us considerable confidence in, for example, eclipses predicted decades or centuries in the future. The solar system is unusual, perhaps verging on unique, in being a naturally occurring system susceptible to analysis of this precision. One reason for this, of course, is that its environment, in so far as this is relevant to its behaviour, is extremely stable.[8]

Of equal or even greater importance in science are concrete

---

[8] For this reason Elizabeth Anscombe (1971) once suggested that 'the success of Newton's astronomy was in one way an intellectual disaster'. The unusual fact that Newton's mechanics had a good naturally occurring model in the solar system led people to embrace a quite unrealistic ideal of scientific explanation

instantiations of models, of two main kinds. First there are artefacts, machines, designed to behave in highly predictable ways in large part through the taking of elaborate precautions to prevent interference with the intended behaviour. Similar in many ways to machines are scientific experiments. Contrary to the impression sometimes conveyed by cleaned-up accounts of experiments, real history of science shows that it is very hard to make experiments work. (I suppose this is familiar to any high school science student.) One of the main reasons for this is, again, that great care must be taken to prevent the interference of factors not considered in the original conception of the experiment. Typically, an experiment will not work when first attempted, and insight and ingenuity are required to work out what unanticipated factors might be interfering, and how they might be neutralized.[9]

Having, finally, got the machine or the experiment to work, this success is conceived as revealing the behaviour and interactions of objects with specific known properties. This is the aspect of the functioning of science that so impresses the casual and the not-so-casual observer. It is often supposed that the experiment reveals these intrinsic properties, and the designer of machines then exploits them in constructing useful objects. In reality the sequence may equally go in the other direction: laws of nature are formulated to explain the workings of machines. At any rate, the processes of building a machine and getting an experiment to work are, as just indicated, closely similar. But as well as allowing the construction of machines or enabling the prediction of the results of experiments, science is generally supposed to be capable of revealing the properties of naturally occurring things in the world, and enabling us to understand and predict their behaviour. If, as seventeenth-century thinkers moved by visions of a divine mechanic supposed, the world consisted of interlocking and nested sets of machines, then this would present no problem in principle. But there is no such mechanic, and the world is not so composed of machines. So a fundamental question in discerning the scope and limits of science based on the kind of procedure just sketched is to what extent the world, or important parts of it, approximates mechanism.

Scientists studying human beings or human societies very often claim to work by constructing models. A model can be seen as the

---

[9] See Galison (1987) for a compelling account of this process in physics.

theoretical analogue of a machine. Where the builders of a machine must take great pains to prevent other factors interfering, however, the builders of a model can achieve this result by fiat: a model considers how some particular set of factors will interact on the assumption that no other factors are relevant. Of course, if other factors are relevant the model will do a proportionately poor job of corresponding to reality. When models perform this job badly the normal response is to supplement them by attempting to incorporate further factors. Sometimes, as for example in the case of contemporary models for economic forecasting, this leads to the production of very complex models indeed, though notoriously great complexity doesn't necessarily lead to great reliability.

There is an assumption behind this process, sometimes explicit, that a sufficiently complex model will accurately represent the system that is being modelled. This is, in essence, the assumption that the systems we attempt to model are machinelike. This assumption, however, is open to at least two philosophical objections. The first of these is related to the hoary philosophical question of determinism. Determinism assumes that any set of causes must interact in a fully deterministic way. The simplest paradigm for such summation of causes is the law of composition of forces. As Nancy Cartwright (1983; 1999) has argued, simple physical laws, such as the law of universal gravitation, are literally false. Many, perhaps all, actual objects are subject to forces other than gravitation, and therefore do not behave in accordance with the law of gravitation. The law of composition of forces, of course, allows us in principle to take account of many physical forces in predicting the motion of a physical object. So physics can tell us what an object will do when subject to a certain gravitational force and a certain electro-magnetic force. On the other hand, physics has no resources for incorporating the further fact that I decide to swallow the object. That there are invariably ways of combining any set of causal factors into a determinate causal upshot is a hypothesis no better established than the controversial thesis of determinism.[10] The only plausible ground for this latter thesis is the also controversial thesis that ultimately there are only physical causes.

---

[10] Or at any rate what I shall later describe, to take account of certain restrictive kinds of indeterminism, as 'causal completeness'. These matters will be discussed in greater detail in Chapter 7.

The second objection relates to my earlier point about machines, that they are conceived in terms of the interactions of intrinsic properties of their constituent parts. Of course, machines are designed to respond differentially to features of their environments. But they are designed to respond to fairly specific and precisely specified features of the environment. In some cases this requires that a suitable environment be constructed, as with the elaborate road systems we construct so as to be able to use motor vehicles. (Compare the way a laboratory is constructed to provide a precisely controllable environment.) Here it is possible to claim a profound disanalogy with the naturally occurring systems, human and otherwise, that are the subjects of the scientific models under present consideration. These latter are dynamically evolving systems in which the environment and the system singled out for modelling develop in multiply interconnected ways. In the course of this book I shall argue that this situation provides important limits to the possibility of understanding them through the kind of methodology just outlined. The fact that certain systems can be shown to respond in highly predictable ways to certain very specific environments is a poor basis for the assumption that any system will respond in a predictable way to any environment. And it is naturally those pairs of systems and environments that do generate predictable outcomes that have particularly attracted the attention of scientists.

There is a rather different attitude to models that should also be mentioned here. It may be suggested that models should not be expected to represent reality even approximately. They rather reveal to us only an aspect of reality, some part of the total, and perhaps cognitively unmanageable, set of things that are going on. Hence we should not be surprised at the notorious failure of many or most economic models, for example, to correspond with reality. Their job is only to provide some insight into reality by indicating a coherent part of it, and perhaps also to tell us what would be happening if we somehow managed to eliminate all the other processes that were happening at the same time. (The utility of this alleged information is an obviously tricky matter.) Much of this book may be seen as indirectly critical of this picture in that I shall try to show the deep deficiencies of the kinds of models that are driven, usually by empirical failure, to appeal to such a rationale. For now I shall mention just two general worries. First, such a conception of scientific modelling is inimical to the robust empiricism I have already endorsed.

There is no problem in supposing that such models are merely a step on the way to the construction of more complex and empirically supported models. But if no such development is predicted or intended, this rationale fails to satisfy the demands of a properly empirical attitude to science. Second, following Cartwright (1983), I see no reason to assume that there must in principle be a way of integrating the processes allegedly disclosed by a model with the further processes allegedly obscuring its influence. So the assumption that in principle, at least, a partial model must be refinable to provide an empirically adequate model seems to me ungrounded.

The greatest defect of reductionism conceived as a regulative ideal, however, is that it has practical consequences for what kind of scientific projects should be promoted. For the projects it speaks for are often highly problematic, and often highly problematic precisely because of their inappropriate emphasis on internal over contextual factors. Some of the most troubling examples are to be found in the medical sciences, and I shall briefly offer just one example here. Consider the several million American children (mostly boys) recently discovered to be suffering from Attention Deficit Hyperactivity Disorder (ADHD) but now being treated with apparent success with the drug Ritalin. According to the American National Institutes of Health consensus statement on the topic, somewhere between 3 and 5 per cent of school-age children suffer from this problem (National Institutes of Health, 1998). Given this prevalence, it is somewhat surprising that such a widespread disorder should have been unknown a few decades ago. But of course this doesn't mean that there were not numerous sufferers. Albert Einstein is sometimes mentioned as a prominent victim.

I do not find it a bit surprising that many children now, and in the past, have had difficulty paying attention in schools. What I worry about is the basis for the conclusion that there is something disordered in the heads of these children which is appropriately treated with psychotropic and addictive drugs.[11] Schools are, after all, often

---

[11] A number of stimulants, including dextroamphetamine, have been found effective in treating these diseases. If a member of the general public were discovered at the school gate dispensing this substance to schoolchildren he or she should anticipate spending many years or decades in jail, not to mention a level of public revulsion exceeded only by that reserved for the sexual exploiters of children. The National Institutes of Health document (1998) notes, not entirely reassuringly, that 'there is no conclusive evidence that careful therapeutic use is harmful'.

boring. It would be instructive to compare the amount of research that has been done on the effects of different teaching techniques on this syndrome, or its effects on the ability to learn different subjects, with that devoted to pharmacological remedies. The fact that powerful drugs can alleviate the manifestations of the syndrome shows very little. One can easily imagine, for instance, that threats of violence would concentrate the minds of recalcitrant students. But this hardly shows that they are suffering from corporal punishment deficiency syndrome. There are many ways of influencing behaviour. My point is not to deny (or affirm) that these children suffer from some common condition that makes them respond poorly to schools; it is to note the ease with which this leads us to assume a medical problem amenable to pharmacological (or not long ago surgical) response. The crucial distinction is that between policies that perceive merely a mismatch between the dispositions of the individual and the social context in which that individual is placed, and those that see inappropriate behaviour as indicative of a deficiency in the individual calling for physical or chemical alteration of the individual. And my point is that it is the latter that reductionist science advertises as possible and desirable. But drugged children, or—to take a different case—slashed, burned, and poisoned patients in unrelievedly carcinogenic environments, are, arguably, the price we pay for action on the basis of the reductionist myth.[12]

This brings us back to evolutionary psychology. I just said that machines are designed to respond differentially to precise and specific features of the environment. But this is exactly what evolutionary psychologists claim of organisms in general, and humans in particular. Hence the mechanistic, reductionistic view of science naturally supports the project of modelling humans as machines with distinguishable mechanical subunits ('mental modules') designed to respond to particular features of the environment. This will be seen, in fact, merely as the application to humans of the mechanistic modelling characteristic of our most successful science.

In the next three chapters I attempt to show why evolutionary psychology is, nevertheless, a largely bankrupt approach to understanding human behaviour. The first of these lays out the main

---

[12] Kaplan (2000) makes this point in detail for a number of cases in medical and behavioural genetics.

difficulties in a very general way; the next illustrates the actual failings with a discussion of one particular aspect of behaviour central to evolutionary psychology, namely, sexual behaviour. The third of these chapters draws some general conclusions, first as to the way that traditional mechanistic model-building is bound to prove inadequate to the study of human behaviour, and second concerning the difficulty of fully separating factual from evaluative issues in an inquiry of this sort. This last is a philosophical issue that is attracting increasing attention from philosophers of science. The sharp separation of factual from normative issues was one of the foundational ideas of classical positivism, but it is increasingly being perceived as untenable.

One of the most important things these mechanistic models prove unable to do is to take sufficient account of the diversity of human behaviour. In Chapter 5, therefore, I attempt to say something more constructive about the nature, extent, and value of human diversity. This last topic, the value of diversity, also provides an occasion to confront the ideological scientism propounded by some of the shriller participants in the so-called 'Science Wars'. In Chapter 6 I turn to approaches analogous in some interesting respects to evolutionary psychology, namely, rational choice theory and its most successful incarnation, neoclassical economics. This chapter provides the occasion for developing two themes that figure less systematically in the preceding chapters, one negative and one positive. The negative one is what I refer to as 'scientific imperialism', the positive one is epistemological pluralism.

By scientific imperialism, I mean the tendency for a successful scientific idea to be applied far beyond its original home, and generally with decreasing success the more its application is expanded. I suspect that this is closely connected with the pathologies discussed earlier. For if one thinks of a subject of scientific study as exhaustively described by an account of its structure, then once one has a good scientific account of something (say, a human being) one should be able to apply it whenever or wherever that thing is encountered. If, on the other hand, one recognizes that context is as important as structure, then such relocations of scientific accounts into new contexts is doomed to failure. This expectation is fully borne out by the examples I consider.

Evolutionary psychology can be seen as a failed imperialistic adventure from evolutionary biology. Imperialistic economics is a deliberate attempt to import the perspective of economics into many diverse facets of human behaviour. Whereas I have no doubt that evolutionary biology, in its place, is a Good Thing, I have serious reservations about neoclassical economics. However, for present purposes I shall not question the appropriateness of this programme for addressing its central questions, questions about the production and distribution of commodities, but restrict my criticisms to its imperialistic exploits. These, it will turn out, have much in common with the deficiencies of evolutionary psychology.

The dangers of this kind of imperialistic use of scientific ideas leads me to the more constructive theme that is especially developed in Chapter 6, the necessity of a pluralistic approach especially to the study of human behaviour. This necessity is illustrated specifically with respect to the concept of work. Economists often treat work as a fairly simple concept, a commodity sold in the labour market by some and used as an input to production by others. But, as I show in this chapter, work carries a considerable diversity of distinct and overlapping meanings in contemporary societies. The point of distinguishing these various and diverse meanings is not, as should by now be clear, to pave the way for a more complex and realistic scientific model. The suggestion is rather that a proper understanding of such a central feature of human life cannot be gained from *any* unique and homogeneous perspective. Neither, therefore, can the application of an adequately sophisticated understanding of such a topic to public policy be grounded in any such unique perspective. The moral of pluralism, therefore, is not to point to better ways of doing science (though certainly it is to be hoped that it will contribute to that too), but to show the limits of science, at least in its application to the complexities of human life.

In the concluding chapter, I show how the pluralistic metaphysics that I advocate in replacement of the mechanistic and deterministic assumptions of much contemporary science and philosophy provides the key to understanding the genuine autonomy of much human behaviour. This gives us one final ingredient of an adequately pluralistic account of human behaviour. We are, as evolutionary psychologists insist, animals that evolved by natural selection; we are also creatures that apply our intellect to choosing among alternatives that life presents to us, as rational choice theorists have described and analysed. But we are also choosers at a higher level, choosers among the plans, projects, and principles that guide our more practical deliberations. These latter decisions determine what sort of difference we make in the world and are, if not uniquely human, at least developed in humans to an unprecedented degree. Here, I argue, we can find what is important about the human autonomy so widely perceived as missing from the more mechanistic perspectives that are the subject of the bulk of this book.

The sceptic about human freedom will naturally want to know where these principles come from that make possible human autonomy. My answer, an answer made possible by the pluralistic metaphysics that underlies the pluralistic methodology just described, is that genuine autonomy resides in the interaction between individuals and society. As I argue earlier in the book, distinctively human capacities derive from the social context in which human individuals are embedded, and most especially, as I stress at several points, from the linguistic resources of a community. It is the exercise of such capacities which constitutes human autonomy. But these socially constituted resources do not determine the behaviour of individuals, but make a range of behaviour possible. And by exercising the capacities for action and argument that the social context provides, individuals can to some degree affect the resources for further action available to themselves and other members of a society. It is in this dialectical relation between the social and the individual that real human autonomy resides. And it is a location that is necessarily invisible to the mechanistic, individualistic approaches against which the polemical parts of the book are directed.

2

The Foundations of
Evolutionary Psychology

1. Introduction

I take it as beyond serious dispute that humans evolved. We share common ancestors with all or most of the other organisms on the planet, and quite recent common ancestors with chimpanzees and other apes. It is also extremely plausible that the main force behind evolutionary change is natural selection, the differential survival of organisms that were better equipped to deal with their environments and leave descendants.

It is a commonplace that these insights have radically changed our view of ourselves. Despite the brilliant arguments of David Hume (1779), prior to the development of a plausible story of how we came to be here it was not unreasonable to suppose that our existence should be credited to the work of some very powerful intelligent being. Now there is little reason to believe anything of that kind, and we must come to terms with the view of ourselves as the product of natural processes. But although, therefore, it is hard to overstate the importance of Darwin's theory to our general view of our place in the order of things, it is easy to overstate the extent to which this theory can generate detailed insight into our nature. Indeed, such overstatement has been a feature of intellectual life ever since Darwin's theory was first propounded (though Darwin himself was very little prone to this error). This chapter will provide an overview of recent attempts to deploy Darwinian theory for the illumination of human nature, and introduce some of the greatest obstacles such a programme faces.

Modern speculations about the implications of evolutionary theory to human nature are conveniently dated from the publication of E. O. Wilson's notorious book, *Sociobiology: The New Synthesis*, in 1975. This book initiated some extremely violent controversy and by the early eighties devastating critiques of Wilson's and related projects began to appear. The most central point to these critiques was the observation that human sociobiology was an almost wholly a priori project. Typical sociobiological arguments turned on the observation that certain kinds of behaviour would have enhanced the reproductive prospects of our ancestors and therefore could be expected to have evolved. The question whether they had evolved was given much less attention. If they were not universally observed, it could merely be remarked that tendencies of the relevant kinds would surely have evolved, but that perhaps they were more or less effectively suppressed by counteracting factors either biological or social. But it was clear all along to evolutionists that the fact that some trait would be useful could not possibly guarantee that it would evolve.

A milestone in the critical response to sociobiology was a seminal paper by Stephen Jay Gould and Richard Lewontin (1979). Gould and Lewontin's paper is actually addressed more broadly to what they called adaptationism, the idea that any arbitrarily selected feature of an organism must be an adaptation, that is to say, must have some function that explains its selection over evolutionary time. With their now famous example of the spandrels on the arches of the cathedral of San Marco in Venice, admirably used as a site for allegorical mosaics, but there solely as an inevitable side-effect of the construction of the arch, they pointed out that in an object as functionally integrated as an organism, particular features will be massively constrained by the overall structure and functioning of the organism.[1] These constraints—developmental, genetic, morphological, and so on—may very well be more important than the optimizing tendencies of natural selection in explaining the final structure of the organism. These criticisms reached their peak with Philip Kitcher's *Vaulting Ambition* (1985), in which the arguments of most of the leading sociobiologists were analysed in considerable detail, and found in all cases to rest on unargued and often quite implausible assumptions.[2]

[1] Inevitably, given the vast amount of discussion this article has generated, it now appears that the question of spandrels was a lot more complicated than Gould and Lewontin acknowledged. There are, for example, various possible solutions to the architectural problem that gives rise to spandrels. For discussion see Pigliucci and Kaplan (2000); Dennett (1995: 267–82).

[2] A more polemical attack was by Lewontin, Rose, and Kamin (1984). The arguments about sexual behaviour were effectively dismantled by Anne Fausto-Sterling (1985).

The response of sociobiologists to this and other criticisms was not so much to address the difficulties pointed out to them, but to disappear and then reappear again under a different guise. In the mid-1980s, sociobiology became an increasingly unpopular title and evolutionary speculations about human behaviour began to appear instead under the rubric of evolutionary psychology. By the end of the twentieth century, evolutionary psychology had established itself as a thriving area of academic life with the full panoply of journals, professional societies, research grants, and conferences. It is true that evolutionary psychologists have begun to pay more attention to empirical information about actual human behaviour. But the main way they have attempted to distinguish themselves from their some- what discredited sociobiological predecessors is in terms of some- what greater clarity as to what is actually involved in the evolution of behaviour.

2. Overview of the Evolutionary Argument

The argument for an evolutionary approach to psychology begins with some apparently familiar and even uncontroversial ideas. First, it is assumed that the way to understand human behaviour is to understand the structure of human brains. But the structure of human brains, it is then said, can be understood at the most funda- mental level by considering the genes, or the genetic programme, that guide the development of brains. And, lastly, the explanation of humans having a particular complement of genes, or a particular genetic programme, is to be sought in reflections on the process of evolution by natural selection. Recent history, it is generally sup- posed, is far too short to produce significant changes in the human genome, so to explain why we have the genes we do we must resort to the long tracts of time during which humans were developing their distinctive genetic endowments, and more specifically the char- acteristically hypertrophied cerebral cortices that distinguish them most strikingly from their nearest relatives. This is generally assumed to be the million or so years preceding modern recorded history, the late Stone Age or Pleistocene. This period is referred to by evolution- ary psychologists as the environment of evolutionary adaptation. Its conditions are those to which, they argue, our brains are primarily adapted.

In distinguishing themselves from their sociobiological precursors, evolutionary psychologists suggest that where sociobiologists had tended to talk somewhat sloppily about the evolution of a behaviour, say rape by males, they talk instead about the evolution of psychological modules in males for rape. It is a central plank of their project that the mind is to be seen as consisting of a large number of such modules, each designed to address a particular adaptive problem posed by the environment of evolutionary adaptation. Against the accusation that sociobiologists had been guilty of a kind of biological determinism, they insist that the existence of such modules does not entail that organisms possessing them will produce the behaviour they were designed to generate. The module will, to begin with, only be designed to produce the behaviour in particular circumstances, circumstances that may fail to obtain. And the modules will interact with other modules directing behaviour, and their characteristic outputs may, thereby, be suppressed.[3] This development certainly added some clarity to the sociobiological position, though it is doubtful whether Wilson or anyone else was really in the dark about the mediation of brains between genes and behaviour. It also provided something like a response to the problem about the lack of empirical evidence for many sociobiological claims, since the argument did not require that any particular behaviour should be observed. The negative side of this, of course, is that it becomes somewhat unclear what would actually constitute evidence for the modules postulated by evolutionary psychologists, and extremely unclear what could constitute evidence against them.

The ultimate success or failure of evolutionary psychology must of course be judged on its products. Does it provide us with plausible and illuminating insights into human nature or human behaviour? To address this question, I shall examine a central set of evolutionary psychology's deliverances in the next chapter. In the present chapter, on the other hand, I shall examine the broader motivations of the programme just sketched. Evolutionary psychologists often suggest that even if their results so far are modest and inconclusive, broad theoretical considerations are sufficient to reveal that theirs must be the right approach in the end. My aim will be to show that, despite some initial plausibility, these considerations are at best

---

[3] Though, as I shall emphasize below, the nature of this interaction is a question they seldom attempt to address.

controversial, and perhaps even quite misguided. The point of this will be to show that we should not be surprised at the failure of the approach, documented in the next chapter, to generate much insight.

3. How Much Must Evolution Explain?

Imagine you are an alien watching a cocktail party with a human social scientist. You ask: why are all these people in this room? The reply comes back: the human species evolved a gene for the metabolism of ethanol. They have all come here to consume ethanol, and eventually they will all go home and, more or less successfully, metabolize it. It is of course true that the evolution of this capacity is a necessary condition for cocktail parties as we know them. So we might even say that the evolutionary observation is, in some sense, part of the full explanation of the phenomenon in question. But it is not a very illuminating part. It is thoroughly unilluminating by comparison to the information that these are all members of the Tunbridge Wells Bridge Club, and this is the Annual General Meeting. The point so far is just that everything we do is something that we have evolved the capacity to do. But the existence of such an evolved capacity may be thoroughly uninformative as to why, when, or even whether we actually exercise that capacity.

Let us develop this fantasy a little further. I imagine that the evolutionary function of alcohol metabolism is probably that it prevents one from being poisoned by very ripe fruit. But suppose, for the sake of argument, that it was actually to enable people to go to cocktail parties (or some prehistoric analogue of cocktail parties). Cocktail parties are no doubt good occasions for cementing social alliances and two mildly drunk people may, for all I know, form stronger social bonds than their sober fellows. So now it is not only the case that evolutionary history explains the possibility of cocktail parties, but it is also true that this possibility evolved precisely because going to cocktail parties furthered our ancestors' reproductive interests. Evolution now provides the correct explanation of why there are cocktail parties at all. Nonetheless, for almost any actual situation in which an explanation was requested, it is still the bridge club calendar or some comparable social fact that remains relevant. That there is an explanation in evolutionary terms for the occurrence of a certain kind of human behaviour is far from implying that evolutionary information typically provides the wanted illumination of why people act.

Let me offer one more fantasy, this time intended to illustrate a case in which evolutionary explanation would be maximally illuminating. Imagine that for many years ancestral human populations were the subject of experimentation by aliens with the scientific skills typical of fictional aliens. In one experiment the aliens scatter a few grand pianos around the Pleistocene environment, and also engineer a gene, which they transplant into a few ancestral humans, that facilitates rapid acquisition of a facility for piano playing. More specifically it directs the production of a brain module which, when activated by a few minutes' contact with a piano, subsequently directs expert playing of the piano. The gene is, by the way, engineered on the Y chromosome, and transplanted only into males. When one of these genetically modified humans encounters a piano he rapidly discovers how to make beautiful music on it. This proves to be irresistibly attractive to ancestral females, and these males have great reproductive success. Unfortunately a few millennia later the research team has its funding removed and the pianos are sold off to a promoter of dance bands on a minor planet of Betelgeuse.

Back to the present. An evolutionist wonders why some people have so much more facility for playing the piano than others and also, being a traditional sexist, wonders why all of those are men. He hypothesizes a gene on the Y chromosome, and suggests that this is a gene for musical ability that attracted mates in the Pleistocene. And, as we have seen, he is absolutely right! I want to suggest that the uncontroversially correct nature of this explanation derives from the coincidence of three factors (and these do not include the intervention of aliens). First, the behaviour is directly caused by a specific structure in the brain; second, the structure in the brain is caused by a specific gene (this could, of course, be a set of genes); and third, the genes are there because they were selected for their ability to produce this brain structure. The natural piano players have a piano-playing module in their heads, it is caused by a specific gene, and this gene was the beneficiary of sexual selection. Of course, as the first example illustrates, this would still only answer some of the questions in which we might be interested. It would not tell us, for instance, why a particular piano player prefers Bach to Tchaikovsky, or why some players are more talented than others. And much more interestingly, it would not explain why pianos are often played on raised platforms in front of large groups of silent watchers. (In such situations it is unusual for fertile females to storm the stage offering their reproductive favours to the male pianist.)

Moving now from science fiction by philosophers to science fiction by scientists, consider a real and controversial example that I shall discuss in the next chapter, rape. Obviously, since a substantial number of men commit rape, it is something that they, and no doubt many men who do not rape, have evolved a capacity to do. This banal observation tells us nothing about why some men commit rape while others don't. But suppose that rape is indeed caused by a particular module in the brain. This could be a module possessed by all males but activated only under certain circumstances (in which case the explanation of rape will of course be incomplete), or a module peculiar to rapists. It must be a module produced by some set of genes, and those genes must have been selected in part, at least, because of their tendency to produce rape-generating brain modules. It is important to bear in mind, as the preceding fables should help to illustrate, that even if the capacity to commit rape is an adaptation as some evolutionary psychologists have claimed—even if, that is, the reason men have the capacity to rape is that ancestral men increased their reproductive fitness by exercising this capacity—it is not obvious how much insight into the occurrence of rape in a particular contemporary society all this provides. Still, if we could be confident that people's brains were stuffed with such genetically determined and adaptively generated modules, it would surely be interesting to know about them anyhow, and it would be plausible that such knowledge would contribute something to our understanding of actual behaviour. So prior to showing, in the next chapter, how thin the evidence for particular such claims typically is, it will be helpful first to point out that there are many reasons to question the a priori arguments offered in support of this vision.

4. Atavism

Evolutionary psychologists claim, as I have already remarked, that our minds are adapted to life in the Stone Age. The only mechanism for acquiring adaptations is said to be natural selection, and natural selection can function only by the accumulation of favourable genes. But the process of accumulating favourable genetic mutations sufficient to produce complex biological structures such as information-processing modules in the brain is a very slow one, and requires much longer periods of time than are provided by the few millennia of recorded history. To find a suitable period of time we must go back to the late Stone Age when, allegedly, humans evolved in a largely stable environment for perhaps more than a million years. It is thus to this environment that we are cognitively adapted. Or so the story goes.

I do not propose to dispute that our genomes evolved over very long periods of time (parts of the genome, indeed, are essentially identical to those of bacteria from which we diverged phylogenetically billions rather than millions of years ago), and that the genetic basis for our brains has probably not changed a great deal since the late Stone Age. Nevertheless, I do want to argue that the centrality of genes both to the story of evolution and to the story of development (the building of brains) is typically greatly exaggerated, and that a proper appreciation of the role of genes in biology does not necessitate the atavistic view of human cognitive adaptation that is central to evolutionary psychological arguments.

Before going any further I must make a crucial terminological point, without which much of what I say throughout this book is likely to be fundamentally misunderstood. Evolution is often defined by biologists to mean change over time in the frequency of genes in a population. Given this definition it is simply a matter of logic that evolution is limited to the rates at which genetic change can accumulate. The definition, however, seems to me highly contentious, and is a major impediment to serious discussion of the rate and nature of change in biological, and most especially human, populations. Throughout this book, therefore, I use the term 'evolution' to refer to any change over time in the distribution of properties across a population. This makes it natural (and important) to consider how much of evolution consists of natural selection of genes.

In obvious synergy with this contentious and genocentric definition of evolution is the well-known claim of Richard Dawkins (1976), developing the ideas of G. C. Williams (1966), that ultimately the only things that natural selection can select are genes. Though this thesis is not an essential component of the evolutionary psychological argument, it has been highly influential in promoting the idea of the centrality of the genetic to our understanding of evolution. The assertion that evolution just is change in gene frequency in a population fits easily with the assumption that natural selection is an engine that directs changes in the frequency of genes towards adaptive ends. As a matter of fact I doubt whether change in gene frequency is even a necessary condition for evolutionary change (see Magnus 1998) though no doubt it almost always occurs. But regardless of the defensibility of any of these theoretical claims, defining evolution in terms of gene frequencies will make many of the questions I want to raise unaskable, and I shall adhere here to the much broader definition indicated above.

Philosophers of biology have, as it happens, almost universally rejected Dawkins's genic selectionism.[4] The main reason for this rejection is that genic selection is unable to describe the richness of causal processes in evolution. Selection is often thought of as discriminating between alleles, or alternative forms of the gene, at a particular point, or locus, on the chromosome. Consider, then, an allele that exhibits meiotic drive, an ability to transmit more than the normal 50 per cent of its copies into the sex cells of its carrier. This is of course an enormously powerful selective advantage, and one that will carry its bearer to fixation (that is, it will supplant rival alleles at that locus for all members of the population) very quickly, other things being equal. Known examples of meiotic drive are necessarily examples that have not gone to fixation (since there must be other alleles relative to which they can subvert meiosis) and must therefore be examples where other things are not equal. These are in fact cases in which the gene is highly deleterious to the organism. In such a case we see a balance between selection of genes (meiotic drive is a real case of genic selection) and selection of individual organisms which works to remove carriers of the gene through their deleterious effects. Simply attributing a selective value, a numerical probability of being selected, to the gene inevitably conceals this complex causal structure. Recently there has also been an increasing interest in the possibility of processes of group selection distinct from either gene or individual selection.[5]

---

[4] See Sober and Lewontin (1982); Sober (1984); Lloyd (1988); Dupré (1993a). A partially dissenting view is that of Sterelny and Kitcher (1988), though their defence is of genic selection in a much weaker sense than Dawkins's, and licenses an exclusive focus on genes in only an instrumentalist sense.

[5] This once popular idea fell into disrepute following the attack of Williams (1966) on Wynne-Edwards (1962). However it has recently come back into fashion. For a detailed defence see Sober and Wilson (1998).

Views of genic selectionism easily mutate into more pernicious forms of genocentrism, or even genetic determinism. We see this in the work of the philosopher most visibly enthusiastic about both genic selectionism and evolutionary psychology, Daniel Dennett (1995). Dennett explains some of his views on evolution by appeal to his image of the library of Mendel. Modelled on Borges's famous library of Babel, a library that contains every possible book, however meaningless, the library of Mendel contains every possible genome. Dennett treats this as if it were a representation of every possible organism. But it is nothing of the kind; it is merely a representation of every possible genome. That this is not at all the same thing can easily be seen by reflecting on the possible fates of identical twins (they may as well, but need not, be human). If one twin is raised in dire poverty, put to work at the age of six, and given no significant education, while the other is raised in an affluent home by well-educated parents, the resulting organisms will be physically and mentally very different. The gene selectionist may be inclined to object that these differences are merely transient; only the genes will be passed on to future generations. But this is plainly false. The fortunate twin will pass on material resources and cultural capital that will make it possible for her offspring to develop much as she has, and the same will go for the less fortunate twin.

This kind of reflection leads to the perspective most radically opposed to genocentrism, developmental systems theory (DST).[6] DST emphasizes that from the point of view of the development of an organism, genes are only one of a range of resources that must be deployed. Others include a vast array of extranuclear chemicals and organelles in the maternal cell, the mother's reproductive physiology, a variety of environmental resources either collected directly from the environment, as food and water, or constructed by the parents from environmental resources, and more or less complex patterns of behaviour employed by parents in rearing offspring. These resources are combined to produce iterated cycles of development, and successful development must produce organisms capable of gathering together all the necessary resources for launching the next generation. There is an important distinction to be drawn among alterations in the cycle between those that can, and those that cannot, be

---

[6] The classic source for this is Oyama (1985). See also Griffiths and Gray (1994) for a lucid overview.

transmitted to future developmental cycles, for evolution by natural selection can only work on alterations of the first sort. In this category genes are of course of great importance, but they are by no means unique; and they become increasingly less unique as animals become more cognitively sophisticated and learn to imitate aspects of the parental behaviour that contributed to their own development. Here I make conceptual space for a banality, but a banality that evolutionary psychologists are often disposed to deny or minimize. The human brain develops in ways that are enormously sensitive to the environment in which it develops. Formal and informal learning, imitation of parents and peers, and so on, presumably affect the physiology of the brain. And, of particular relevance to the present argument, the social structures that produce these effects are constantly reproduced in ways that make possible the more or less similar development of subsequent generations of human brains.

In opposition to the almost universal tendency to talk about genes as repositories of information, embodying codes, and so on, DST denies that there is anything unique about the role of genes in either development or evolution. An important recent article by Peter Godfrey-Smith (2000) suggests that there is a legitimate interpretation of genetic coding as referring to the relation between genes and protein chains, but not between genes and any more complex phenotypic features. This strikes me as an important qualification of DST, but in no way one that challenges the central critical argument of that programme.

We can now see how massively simplistic is the assumption that genes build brains. Obviously genes can do nothing, let alone build a brain, on their own. To build a human brain the genes must be properly located in a cell complete with all the properly functioning extranuclear machinery; the cell must be properly positioned in the uterus of a human female; and the child must be born into a social setting that will provide an extremely complex set of stimuli to the human organism in which the brain is located. So is there really any sense in which genes build brains in which it is not equally true to say that wombs build brains, or even schools build brains? Perhaps it might be thought that genes are the real causal agents, while these other things were merely background conditions. This is a problematic philosophical distinction, but one interpretation does suggest itself here. Perhaps a particular set of genes will always build a brain of a certain sort if the additional conditions are adequate. If they are not adequate, these genes will either produce an unequivocally defective brain, or no brain at all. This picture is certainly encouraged by the almost universal tendency to describe genes as blueprints or instructions. Though no one would be tempted to claim that the architect's blueprint built the house, it does have a unique relation to the final house. If all goes well a particular blueprint will always produce a house of a more or less exactly specified kind. The bricks, mortar, carpenters, and so on, which are obviously causally necessary to the appearance of the house, could nevertheless have been deployed in the construction of an indefinite range of possible buildings. The blueprint determines which of these possible buildings actually emerges. Something analogous is, I think, implied by the suggestion that genes build brains.

But the analogy just suggested is entirely broken-backed. Even the most enthusiastic supporters of the architectural virtues of genes know perfectly well that the properties of brains depend on much besides genes, and that this is true even if everything involved in the expression of genes goes well. Supporters of the genetic basis of intelligence, for example, claim typically that 50 per cent of intelligence is genetically caused.[7] But obviously this implies that another 50 per cent is not genetically caused. And no subtle statistical investigations are needed to tell us that well-nourished children raised in intellectually stimulating environments rich in love and attention will generally develop better-functioning brains than those raised in materially, emotionally, and intellectually impoverished environments. All of which simply emphasizes the point that genes are not a blueprint for bodies or brains, but one among a range of causal factors that interact in an exceedingly complex sequence of processes to produce bodies and their brains. This does not lead to what evolutionary psychologists like to parody as the 'Standard Social Sciences

---

[7] See, for instance, Herrnstein and Murray (1994: 107–8). This is a claim of a kind often revealing a great deal of confusion. They are generally based on analysis of variance and the specific conclusion here is that 50 per cent of the variance in intelligence is 'explained' by genetic differences. But as is often pointed out, this is a notion that is itself thoroughly context-dependent. If we succeed in bringing up people in identical environments it will follow as a matter of definition that 100 per cent of the variance in intelligence is explained genetically. It would be strange indeed to conclude that in this case intelligence was 100 per cent genetically caused. Conversely, if we produced a large number of genetically identical clones—a much more feasible if no more morally appetizing experiment—we could discover that for them intelligence was wholly environmentally caused. The confusions in this area are particularly clearly explained by Block (1995).

Model' (Tooby and Cosmides, 1992), the idea that brains are blank slates developing with infinite plasticity in response to environmental variation. What it does imply is that since the conditions under which contemporary brains develop are very different from the conditions under which human brains developed in the Stone Age, there is no reason to suppose that the outcome of that development was even approximately the same then as now. And consequently, if we want to know what contemporary human brains are like, reflection on the conditions under which humans (perhaps) lived in the Stone Age is no substitute for the hard empirical work of investigating the nature and variety of contemporary humans.

Critics of DST complain that it fails to offer any positive programme that has achievements comparable to more orthodox neo-Darwinism, and so far this complaint is probably justified. DST is at any rate much less controversial as providing a critique of genocentrism than as offering such an alternative positive programme. The common suggestion that evolutionary psychology is an inevitable consequence of contemporary neo-Darwinism should be qualified with the realization that its interpretation of neo-Darwinism is a controversial one against which powerful criticisms have been levelled. The alternative to seeing genes as providing a blueprint for the construction of brains is not, as evolutionary psychologists like to maintain, the idea of a blank slate to be written on in any way at all by the environment, but of a brain constructed by a variety of more or less stable and reliable resources including resources that are reliably reproduced by human cultures. And this picture, an extremely convincing one in my view, makes plausible the idea that human brains evolve[8] at the speed of cultural change rather than at the speed of accumulation of genetic change which, finally, makes cognitive atavism entirely optional.

5. Do Brains Cause Behaviour?

Let us suppose, despite the foregoing discussion, that the structure of the brain is largely determined by a million-year-old genetic

---

[8] This might be a good place to remind the reader that I am using the term 'evolution' simply to mean change in the features found in a population, and hence in a sense that does not beg the question whether all evolution is genetic change.

programme. We still need to ask to what extent the deciphering of this programme and its implicit records of Stone Age life will enable us to explain or predict the behaviour of humans. Although I suspect that many contemporary thinkers will take the answer to the question at the head of this section as obviously affirmative, I want to suggest that matters are rather more complicated.

It will be helpful to approach the difficulties with the claim that brains cause behaviour by starting with a more traditional philosophical thesis, the thesis that features of the mind explain actions.

This traditional thesis is often associated, and sometimes disparagingly associated, with so-called 'folk psychology'. Central to folk psychology is the way ordinary people explain their and others' behaviour. Typically, the story goes, we explain why someone acts in a certain way by saying what they were hoping to achieve and claiming that they believed the action would achieve this goal. A standard banal example: Why did he take his umbrella? Because he believed that it was going to rain and that his umbrella would prevent him from getting wet, and he wanted to stay dry. This is an example of the so-called 'belief/desire model' of action explanation. I shall say more about this model, some of it critical, in later chapters. But for now I shall take for granted that it is a roughly correct model for the explanation of behaviour. Contemporary scientifically minded philosophers often take this model to be a good starting point. However they typically go on to say that its adequacy depends on the extent to which the beliefs and desires cited by folk psychology can be identified with properly physical states of the brains that are the real physical causes of the bodily movements that constitute the action to be explained. The question that I want to consider is whether the translation between these two pictures is really that simple.

Daniel Dennett, in a characteristically and refreshingly forthright manner, writes: '*Of course* our minds are our brains, and hence are ultimately just stupendously complex "machines"; the difference between us and other animals is one of huge degree, not metaphysical kind' (1995: 370, italics in original). The reference to metaphysical kinds naturally suggests that the contrast is with Cartesian dualism (and indeed Dennett has just made passing mention of Descartes). But the assumption that the only options are ontological dualism and the thesis that our minds are our brains is wholly unwarranted. It is, indeed, a drearily Cartesian assumption, for one of the most baleful aspects of the Cartesian legacy is the assumption that either there is only one kind of thing in the universe, mind or matter, or there are two, mind and matter. A much more interesting and plausible option than either, I suggest, is that there are many kinds of things.[9] There may perhaps be a sense in which there is only one kind of stuff (though I suspect that this common belief depends only on a vacuous definition of what it is to be physical), but the interesting question is then whether all there is to being a metaphysical kind is being composed of a certain kind of stuff; a question that only needs to be raised to be answered in the negative.[10] I shall return to these broader metaphysical issues, and also to the strange legacy of mechanism displayed in the second clause of the quote from Dennett, in the final chapter, but for now I want to consider some respects in which minds, human minds at any rate, differ from brains.

One familiar way in which humans differ in kind from other animals is in their possession of complex languages. Nothing, of course, turns on whether some animals might turn out to have languages more similar to human languages than is generally supposed. If so, they would have whatever other properties follow from having languages of particular kinds. Humans, at least, have complex languages. One way of showing that the mind is not the same thing as the brain is by way of the following two claims: first, that it is impossible to characterize a human mind without appeal to language; and second, that a language is not a property of an individual but of a linguistic community. From these premises it follows that it is not possible to characterize a mind without some reference, implicit or otherwise, to a community of which the possessor of that mind is a member. This argument is not simply about the consequences of describing an object a certain way (as a mind), consequences that might not apply if the object was described from some other perspective (as a brain). The argument contends that aspects of the mind depend ontologically on the community in which they are embedded; they would not be the features they were if they were not related in the right way to the community in which they are

---

[9] The phrase 'kinds of *things*' probably concedes too much. There surely exist minds, but it is probably confused to refer to them as things. A further aspect of the confusion stemming from Descartes and nicely exemplified by Dennett is the idea that a mind is properly said to be made of stuff at all, physical or ghostly. And calling a mind a 'thing' suggests that there must be some stuff or other of which it is composed.

[10] The argument for metaphysical pluralism is developed in detail in Dupré (1993a).

embedded. But it would be absurd to claim that the brain depends ontologically, or constitutively, on its social context, so it follows that the mind cannot be identified with the brain.[11]

The first premise is, I suppose, obvious. Establishing it does not require that I take a position on the somewhat controversial question of whether non-linguistic animals possess mental properties,[12] or whether, as some philosophers hold, a good deal of human thought exists quite independently of real language, formulated in an innate 'language' of thought.[13] The claim is only that there are features of the human mind that could not exist apart from language. Whatever may be thinkable in the language of thought, it is hard to believe that propositions of quantum mechanics or metaphysics, say, could be so thinkable. Such prelinguistic vehicles of thought are conceived as having evolved prior at least to any contemporary languages, and it is difficult to believe that they could have been somehow preadapted to entertain thoughts that were so far beyond the repertoire of their early subjects.

I am inclined to think that this premise, the dependence of much human thought on language, is, like the existence of the external world, more certain than any argument that could be adduced in its favour. In case it is not, however, the following line of thought should provide it with further support.[14] Humans are self-conscious. They not only represent the world to themselves, but they are aware that they do so. This involves awareness of a distinction between the way a person represents the world and the way the world is, and inevitably give rise to the idea that there are different possible representations of the world, and that these may be more or less adequate to it. But this, finally, implies awareness of a realm of representations of the world which for us, at least, is a realm of linguistic entities. We can only be conscious of ourselves as knowers to the extent that

---

[11] This absurdity may not be entirely obvious to those less familiar with philosophical terminology. The brain is, as I have emphasized, causally dependent on features of its social context. But by 'ontological dependence' I mean something much stronger, that it would be logically impossible for the dependent entity to exist without the correct relation to the entity on which it depends. So, for example, money is ontologically dependent on the social consensus that it is, indeed, money (see Searle, 1995). Nothing like this, I take it, is true of the brain, qua physical structure.

[12] Though I have no doubt they do. See Dupré (1990).

[13] Not, I take it, for reasons that will become clearer, a language.

[14] An argument along these lines is interestingly developed by O'Hear (1997).

our knowledge involves symbolic representation. Since we are self-conscious in this way, we must in fact possess knowledge in the form of symbolic representations of the world.

So, turning to the second premise of the argument, why shouldn't the brain somehow contain symbolic representations? The answer to this, though surprisingly little appreciated, has been made perfectly clear by Wittgenstein. It is just the recognition that meaning, that which gives a symbol the capacity to represent, is a normative concept. What makes it correct to apply a symbol to a thing or a situation? If there are no criteria of correctness outside the head of the person using a word, whether using it to speak or to think, then, as Wittgenstein famously remarked, whatever seems right is right (1953: §258). But this, finally, leaves no difference between being right and being wrong. Meaningfulness, therefore, what makes a usage right or wrong, depends on the existence of norms, or rules, which is to say practices with normative force in a community. And therefore the existence of meaningful symbols requires the existence of a community. The brain, or even the mind, can only contain symbolic representations by virtue of the relation of its possessor to a community of symbol users. And hence also, connecting this line of thought with the argument summarized in the preceding paragraph, the possibility of self-consciousness is a possibility only for a being embedded in a linguistic community.

Although this line of argument seems to me impeccable, I know from experience that it often fails to convince. Probably the main reason it fails to convince is that we are still committed to a classical conception of meaning for which the paradigm is direct confrontation with the object to which a word refers. For a classical thinker such as Locke, this is conceived as direct confrontation with an inner object, a sensation, precisely the situation that Wittgenstein is addressing and rejecting with the remark just cited. More recent philosophy has tended to see meaning as derived from confrontations with real things encountered in the world, but the same paradigm of direct confrontation with the referent survives. One of the main objectives of Wittgenstein's *Philosophical Investigations* is precisely to demonstrate that such direct confrontation with an object is wholly inadequate to setting up a relation of meaning between a word and an object: 'one forgets that a great deal of stage-setting in the language is presupposed if the mere act of naming is to make sense' (1953: §257). Wittgenstein's arguments are, however, notoriously difficult and controversial, and it would be beyond the scope of the present work to examine them in any detail. For those who are not persuaded by this Wittgensteinian argument, I offer a rather different line of thought that leads to a similar, but more general, conclusion.

Even apart from issues directly connected with language and thought, I claim that humans must be thought of as ontologically dependent on their social contexts. Consider the capacities possessed by a particular human being. One might first think of a set of 'bare' capacities. By these I mean the purely physical abilities to move one's limbs, produce articulate sounds, and so on. Clearly this provides a very inadequate account of what people can do. We are a long way from the capacities, say, to play golf, to write a cheque, or to run for political office. Quite obviously, capacities of these latter kinds do not depend solely on the entity that displays them being a physical object of a certain kind. The physical object must be suitably related to a social context that makes such things possible. We do not, however, need to go to such a rarefied level to see the context-dependence of human capacities. Think, for instance, of the capacity of most humans to move vertically through tall buildings. For most of us this is achieved by virtue of having functioning legs, and by the provision of staircases. However, for many others who lack functioning legs but possess functioning wheelchairs, it is achieved by the provision of lifts. In a society in which no buildings were equipped with lifts, these latter people would wholly lack the capacity in question. In a society in which all buildings were so equipped, they would have this capacity to just the same extent as those able to use stairs. Nowadays it is legally required in some countries that certain kinds of buildings be equipped with lifts, and the capacities of those confined to wheelchairs are consequently enhanced. This enhancement is acquired through a social decision. It is quite generally true that the ability to move about in modern societies is much more a matter of socially provided resources than it is of bare capacities. For negotiating substantial distances, access to a car or the resources to purchase train or plane tickets are much more significant than is the proper functioning of one's legs.[15]

It will be objected, and rightly, that essentially the same point applies to the capacities of non-human animals. A cat has the cap-

[15] This argument is developed further in Dupré (1998) and is much indebted to Perry et al. (1996).

acity to catch mice, but only in an environment in which mice are there to be caught. Or even more basically, a bird has the capacity to fly only in an environment in which there is an atmosphere with sufficient density to support flight. And of course animals evolve capacities precisely in relation to an environment which makes those capacities simultaneously possible and advantageous for its survival. As is sometimes remarked, the relation between an organism and its environment is an extremely intimate one, and in important respects the animal may actually construct an environment suited to its capacities.[16] Nevertheless all this does point to something, if not unique about the human species, at least so far developed as to make a distinction of degree approach a difference of kind. Humans are social animals, which is to say that the environment for which they are adapted is human society. But the capacity to live in a *human* society is a very special one. Though no doubt human infants are typically born with the capacity to live in any grossly functional human society, they develop an extremely sophisticated and elaborate set of capacities for living in a particular one. They learn the language and customs of the society in which they live. Some humans, no doubt, become more or less cosmopolitan and can function in a variety of cultures. Most readers of this book would probably do well enough in any anglophone culture, and would probably manage more or less well in most European cultures. Some will perhaps be anthropologists who have learned to live among hunter-gatherers or Islamic fundamentalists. But most of us would be wholly at sea in those environments.

The point, hardly an original one, is that it is a central part of our biology to live in complex societies, and these societies vary in very significant ways. Familiar though this point may be, it is one that seems constantly to be lost on those who wish to provide universalizing scientific accounts of human nature. It is considerations of these kinds, rather than a naive appeal to blank-slateism as often alleged by evolutionary psychologists, that give real philosophical purchase to the complaint that reductionist scientism fails to take proper account of the cultural determinants of human behaviour. Evolutionary psychologists, unsurprisingly, do not merely miss the point, but in certain respects attempt to deny it, or at least to deny

[16] For a sophisticated account of the relationship between evolving organisms and their environment, see Brandon (1990).

that these societies differ from one another in significant ways. In fact they will surely have identified the last few paragraphs as one more tired restatement of the Standard Social Sciences Model, the model according to which humans are equipped with a wholly general information-processing mechanism that allows them to develop with infinite flexibility to meet the demands of whatever culture they happen to be in.[17] At this point I cannot avoid a rather more direct confrontation with the hoary issue of nature and nurture.

6. Nature and Culture

One of the many curious features of the interminable debates over nature and nurture is that when disputants on either side of the issue retreat to more theoretical discussions, they tend to say more or less exactly the same thing. The invariable strategy is to insist that *of course* all interesting human behaviour is caused by the interaction of biological and environmental, often cultural, factors, and it is therefore unintelligible that their benighted opponents continue to maintain that some or all of it is purely biological or purely cultural. So, for example, John Tooby and Leda Cosmides, in what has come to be seen as a classic theoretical defence of evolutionary psychology, write the following:

[T]he whole incoherent opposition between socially determined (or culturally determined) phenomena and biologically determined (or genetically determined) phenomena should be consigned to the dustbin of history. (1992: 46)

Similarly Stephen Jay Gould, in an article Tooby and Cosmides cite as a defence of the Standard Social Sciences Model, writes:

For Linnaeus, *Homo sapiens* was both special and not special. Unfortunately . . . [s]pecial and not special have come to mean nonbiological

---

[17] An interesting, if speculative, argument is relevant here. Sober (1994: ch. 3) has suggested that in a stable environment it might be better to have behavioural responses hardwired as this will save on the costs and risks of learning behaviour later in the developmental cycle. In a rapidly changing environment, on the other hand, the potential for more rapid changes in behaviour across generations will make learning behaviour more advantageous. If this argument is right, it highlights the importance to the evolutionary psychological argument of the questionable assumption that conditions in the Stone Age were largely unchanging for long periods of time.

and biological, or nature and nurture. These later polarizations are nonsensical. Humans are animals and everything we do lies within our biological potential. (1977: 251)

The title of Gould's essay is 'Biological Potential vs. Biological Determinism', so presumably we should emphasize the last word in the above quote. But biological determinism is something that Tooby and Cosmides not only see as incoherent, but also strenuously deny. And indeed they claim that 'it is a complete misperception to think that an adaptationist perspective denies or in the least minimizes the role of the environment in human development, psychology, behavior, or social life' (1992: 87). And even E. O. Wilson, to whom Gould's essay is less anachronistically addressed, considers behavioural flexibility one of the things that are genetically determined: 'genes promoting flexibility in social behaviour are strongly selected at the individual level' (1975a: 549).

Wilson certainly talked freely about genes for this or that quite specific behaviour but, as I have noted, contemporary evolutionary psychologists have distanced themselves from this kind of talk, and speak only of genetic determination of mental mechanisms. How much difference does this make? Consider one example. Daly and Wilson (1988), in a paper Tooby and Cosmides cite with approval, describe in some detail what they take to be the evolved mental module which relates to men's attitudes to women. This includes such things as a tendency to reject raped women as 'damaged goods', a tendency to proportion parental care to degree of confidence in paternity, and a tendency to see wives as a form of property. The evidence for these features of the module, apart from a priori biological argument, are claims about the behaviour of men in various societies, and about norms or laws in various societies relating to sexual relations between men and women. Clearly these are only evidence for features of the module if the existence of a module with such features will at least increase the probability of such behaviour or of the existence of such rules. Since it is also supposed that there are genes whose function is to contribute to the construction of mental modules, it surely follows that there are genes that raise the probability of such behaviour or social rules. And since raising the probability of a trait or behaviour, x, is the most any careful practitioner will allow as a legitimate interpretation of the phrase 'a gene for x' (see e.g. Dawkins 1982: 21), it seems clear that contemporary evolutionary psychologists do in fact still think that there are genes for rejecting raped women, treating women as property, and so on.[18]

These are, in fact, examples of what Tooby and Cosmides call cultural universals. It is such universals that 'incoherent' culturalists refuse to admit, and that it is a central part of the task of a properly reformed psychology to discover. And there are a good number of them:

adults have children; humans have a species-typical body form; humans have characteristic emotions; humans move through a life history cued by observable body changes; humans come in two sexes; they eat food and are motivated to seek it when they lack it; humans are born and eventually die; they are related through sexual reproduction and through chains of descent; they turn their eyes towards objects and events that tend to be informative about adaptively consequential issues; they often compete, contend, or fight over limited social or subsistence resources; they express fear and avoidance of dangers; they preferentially associate with mates, children, and other kin; they create and maintain enduring, mutually beneficial individuated relationships with nonrelatives; they speak; they create and participate in coalitions; they desire, plan, deceive, love, gaze, envy, get ill, have sex, play, can be be injured, are satiated; and on and on. Our immensely elaborate species-typical physiological and psychological architectures not only constitute regularities in themselves but they impose within and across cultures all kinds of regularities on human life, as do the common features of the environments we inhabit. (Tooby and Cosmides, 1992: 89)

The fact that most of this list is quite banal is crucial to its highly tendentious function. It does correctly point out that humans have a rich and complex biology, and that this constrains the kinds of lives we can lead. Gould makes the opposite rhetorical move when he writes: 'If genes only specify that we are large enough to live in a world of gravitational forces, need to rest our bodies by sleeping, and do not photosynthesize, then the realm of genetic determinism will be relatively uninspiring' (1977: 253). True enough, but as Tooby and Cosmides's list illustrates, there is a bit more to it than that. The tendentious aspect of that list is that lurking among the banalities of eating, speaking, having sex, etc., are some strikingly theory-laden Trojan horses.

It is, for example, banal that humans turn their eyes towards

---

[18] For a useful discussion of the use of the locution 'gene for x', see Kaplan and Pigliucci (2001).

things, but that these are generally things that are 'informative about adaptively consequential issues' is not a thesis for which any evidence is offered. I suspect that many people walking down typical high streets have no particular tendency to turn their eyes towards super-market windows, despite the fact that they are full of adaptively consequential things (food), but are more likely to have their gaze drawn to displays of fashionable clothes, sports equipment, toys, electronic equipment, or whatever else may be among their particular not very adaptive interests. The tendency to associate with non-immediate family is surely a declining one in modern Western societies and I suspect, though I offer no more evidence here than do Tooby and Cosmides, that increasingly many people associate more with friends formed on the basis of common interests than with relatives. What is surely true is that apart from the benefits of friendship itself (a thankfully positive sum game), relationships with non-relatives are not typically conceived as based on mutual benefit.

When we get a list of some of the mental modules that underlie the less purely physiological universals a bit later on, it is clear that the Trojan horses have been breeding like rabbits:

a face recognition module, a spatial relations module, a rigid object mechanics module, a tool-use module, a fear module, a social-exchange module, an emotion-perception module, a kin-oriented motivation module, an effort allocation and recalibration module, a child-care module, a social-inference module, a sexual-attraction module, a semantic-inference module, a friendship module, a grammar acquisition module, a communication-pragmatics module, a theory of mind module, and so on. (Tooby and Cosmides, 1992: 113)

The only point that needs to be made here is that whereas there may be good reasons to think that innate structures in the brain are important for some of these tasks, for others (social exchange modules, or effort allocation and recalibration modules, for instance) this is the wildest speculation. And even where there are such reasons the degree of specialization and isolation from other tasks is again a completely unresolved empirical issue.

Tooby and Cosmides are scornful of the tendency of their opponents to assume that the kind of genetic determination they are advocating implies rigid and unalterable behaviour. Clearly contradicting this assumption, Wilson and Daly (1992) write, for example, 'In some societies nothing is more shameful than to be cuckolded, and a violent reaction is laudable; in others, jealousy is shameful, and its violent expression is criminal' (1992: 313). As they explain shortly afterwards, 'Cultural diversity exists, to be sure, but its rationales will not be understood until the cross-culturally general human nature that enables it is elucidated' (ibid.). But the problem is not that a module designed to elicit violent reactions to cuckolding could not somehow be involved in the evolution of a society in which jealousy is shameful. Perhaps the latter norm was explicitly introduced to reduce the mayhem caused by the mental module, for instance. The problem is that cultural variability makes it difficult or impossible to provide evidence for these modules. When we move away from the abstract theoretical discussions of the general desirability of a highly modular mind on alleged engineering grounds, and look at specific claims about the kinds of modules that actually exist, we find only two lines of argument. One is reflection on the problems facing our ancestors in the Pleistocene, a strategy which, for various reasons discussed in this chapter, is fraught with problems and dangers, and the other is to examine statistically the behaviour of living human beings. But once we accept the evolutionary psychologists' defence that the existence of modules doesn't imply the universal production of behaviour of the kind the modules were designed to produce, it is quite unclear how even a significant preponderance of behaviour in accordance with the supposed design of the module would provide convincing evidence for the existence of the module. And that is true even if we accept the highly debatable arguments for the existence of such domain-specific modules in the first place. So short of advances in neurobiology that enable us to demonstrate the function of bits of brains directly, it is quite obscure how the programme is to proceed at all.

There are a number of other general issues that might be raised concerning the foundations of evolutionary psychology. For example, there is an issue whether arguments drawn from the demands of Stone Age life do not appeal irresponsibly to the assumption that evolution will somehow produce the best possible response to environmental conditions rather than merely the best available. Without this assumption it is difficult to see how any conclusions can be drawn from such speculations, since we certainly have no detailed knowledge concerning the range of genetic variation and constraints encountered by our distant ancestors. But the assumption of optimality is an untenable one in contemporary evolutionary theory.[19] And untenable assumptions of optimality to one side, do we really know enough about Pleistocene conditions to license inferences from these to what must or might have evolved?

Another set of problematic issues concerns modularity. If the mind consists of hundreds or thousands of specialized modules, who decides which one should take precedence at a particular moment or how their inputs are to be integrated? The spectre of a homunculus begins to loom. I touch again on this problem in the next chapter.

Still, there may remain a reasonable worry that these objections as well as those that have been discussed in more detail in this chapter have remained at a rather abstract level, and in science the proof of the pudding should be in the eating. I have not tried to demonstrate that evolutionary psychology is incoherent, or self-evidently false, only that the grounding assumptions by which it is motivated are far more problematic and controversial than its practitioners generally allow. Once again, we should turn to the pudding. If evolutionary psychology is really increasing our understanding of the human mind, perhaps only by generating hypotheses that turn out more often than not to be true and perhaps even unexpected, maybe we should not worry too much about its dubious foundations. In the next chapter, therefore, I shall look in more detail at some of the more specific claims generated by evolutionary psychologists, particularly in the area to which probably the greatest amount of effort has been devoted, sexual behaviour. I shall conclude, however, that these practical results are no more impressive than their conceptual foundations. Tasting the pudding, in the end, leads to a serious case of indigestion.

[19] For detailed discussion of issues concerning the role of optimality assumptions in evolution see the essays collected in Dupré, 1987a. A flavour of the problem can be given with the following little anecdote, for which I am indebted to Patrick Byrne. Two anglers, so the story goes, are fishing for salmon in Alaska. Suddenly one of them sees a polar bear rapidly approaching. He hurriedly removes his waders and begins putting on his sneakers. 'You fool,' says his companion, observing this activity and its cause. 'You can't outrun a bear whatever you're wearing.' '*You* fool,' says the first. 'I don't have to outrun the bear, I only have to outrun you.' The point of this little episode of natural selection hardly needs elaboration. Survival of the fittest, as noted in the text, doesn't mean survival of the fittest possible, just survival of the fittest actual. Design analysis of hypothetical ancestral organisms has little significance unless we know a good deal about the range of behaviour actually exhibited by actual ancestors.

3

The Evolutionary Psychology of Sex and Gender

1. Introduction

The aim of this chapter is to engage in some detail with the nitty-gritty of contemporary evolutionary psychology, especially the evolutionary psychology of sex and gender. It will be helpful to begin with a brief consideration of the distinction just invoked, that between sex and gender. The distinction originates in feminist scholarship with the insistence that gender, the differentiated roles and identities defined for men and women by particular cultures, should be sharply distinguished from sex, the supposedly universal biological differences between men and women. The central claim was simply that sex did not determine gender roles. The support for this claim was a wide variety of empirical investigations of the variability of gender roles both cross-culturally and through human history. Since it was generally assumed that biology was more or less a constant across these diverse contexts,[1] this diversity seemed to show that sex did not determine gender. This led to a positive concern with how gender roles were shaped and maintained, and a political engagement with the question how they might be changed.

As I have noted in the previous chapter, contemporary evolutionary psychologists generally acknowledge some degree of variation among human cultures. As I have also suggested, this acknowledgement is not without its problems, most notably the difficulty it presents in providing empirical support for their hypotheses. And in

---

[1] Some feminists later came to question even this assumption, and recognize a relation of mutual determination between sex and gender (see Jaggar, 1983: 109–13). In the 1990s, feminists began to argue that sex was just as much a social construct as gender (e.g. Butler, 1990). Though interesting and important, these developments do not materially affect my present points.

fact it is highly characteristic of evolutionary psychology to insist that the extent of diversity has been greatly exaggerated by anthropologists labouring under the illusions of the Standard Social Sciences Model. They delight, for example, in citing Freeman's (1983) claim to have refuted the classic ethnography of Samoa by Margaret Mead (1949), the latter having been the *locus classicus* for claims about the variability of human sexual behaviour. Evolutionary psychologists, in short, admit that variability exists on pain of empirical absurdity, but deny that there is nearly as much of it as their opponents claim.

I do not propose to attempt to adjudicate the question exactly how much variation in gender roles there may be. Fortunately it is admitted on all sides that there is a good deal of it, and this will be sufficient for the purposes of the present discussion. Evolutionary psychologists want to claim, nonetheless, that the key to understanding the various manifestations of gender in human societies is to expose the species-wide psychology of sex on which these various structures are all erected. And it is to this project that I now turn.

2. The Sociobiology of Sex and Gender: The Classic Story

The starting point for all sociobiological stories about sex and gender is with what is now taken to be the fundamental biological definition of male and female. In sexual species there is generally a large disparity between the size of the gametes (sperms and eggs) that unite to form the zygote which, in turn, develops into a new organism. Males, by definition, are the contributors of the smaller gamete, females of the larger. Introducing an economic metaphor, to which I shall return, males are said to require a much smaller investment in reproduction.[2] In most animal species, of course, this discrepancy in gamete size is only a tiny part of the difference in biological investment in reproduction: for mammals, in particular, the female contribution also includes gestation and, usually, a substantial amount of post-natal care including lactation.

This difference in investment, the story then goes, will lead males and females to pursue radically different strategies in seeking to

---

[2] This economic conception of the problem was popularized by Trivers (1972).

maximize their reproductive success.[3] Males, whose gametes are cheap and numerous, will seek to mate with as many females as possible. This will lead to various kinds of more or less violent conflict between males over access to females, reluctance to devote much energy to any one female, and, it is often suggested, various deceptive or coercive strategies in seeking matings. As Richard Dawkins puts it, 'a male . . . can never get enough copulations with as many females as possible: the word excess has no meaning for a male' (1976: 176). Females, on the other hand, have their potential for reproduction much more limited by the large investment demanded by each offspring and, given male psychology, experience no difficulty in acquiring the minimal necessary male contribution to the process. They will, therefore, rather be concerned to obtain male mating partners with the highest quality genes and, if possible, to mate with males who are willing to contribute something to the care of the offspring. Since the male, having made his small contribution to mating, has little evolutionary reason for hanging around, it is generally supposed, however, that the latter desideratum is usually unattainable. So far this story is intended to apply quite generally to sexual organisms, though with greatest force to organisms with the most disparity between the reproductive investment of the two sexes. It is also fair to say that it is a story that has provided some insight into the variety of mating behaviour observed in nature.

It is crucial to emphasize, however, the *variety* of such behaviour. There is enormous diversity among species in the degree of promiscuity or monogamy in both sexes, and enormous diversity in the ways in which different animals select their mates. This variability is fully exhibited by our closest non-human relatives. Whereas chimpanzees are highly promiscuous, fertile females generally being observed to mate with several males, and their close relatives the bonobos have become a byword for polymorphous perversity, silverback gorillas, the dominant males, enjoy exclusive access to a group

---

[3] As is customary in evolutionary discussion, I use teleological language such as 'seeking' reproductive success. It should be understood that strictly speaking all that is intended is that ancestral organisms that pursued such strategies were reproductively successful; and their more numerous descendants inherited the tendency to follow these strategies. As this makes clear, such arguments must always assume that some ancestral organisms did indeed pursue the strategy in question, and that the tendency to do so was genetically transmitted. As sociobiological thought becomes more speculative these assumptions become anything but trivial.

of females.[4] Thus it is extremely hazardous to infer what kind of mating behaviour to expect in a species apart from detailed and careful observation of the animals in question. This brings us to the application of all this to humans, and its problems.

Early sociobiologists exhibited varying degrees of caution in the extension of their theories to humans, but some general ideas were widely asserted or insinuated. It was taken as fairly obvious that men are inclined to promiscuity and women to monogamy, and thus that, in the words of one authority, 'In . . . all human societies, copulation is usually a female service or favor' (Symons, 1979: 202). Women, but not men, were assumed to have a biological urge to take care of children, whereas men were expected to be out in the forest—or its modern surrogate, the urban jungle—competing with one another for resources and, ultimately, access to more women. In summary, let me quote Wilson himself:

It pays males to be aggressive, hasty, fickle and undiscriminating. In theory it is more profitable for females to be coy, to hold back until they can identify males with the best genes. In a species that rears young, it is also important for the females to select males who are likely to stay with them after insemination.

Humans obey this biological principle faithfully. (Wilson 1978, 125)

Needless to say, such pronouncements reflected some widely held stereotypes. However, it was also widely perceived that such stories were extremely simplistic. The evidence on which they were based was often little more than the stereotypic impressions of sociobiologists, and little account was taken of the huge variety of human sexual behaviour, let alone variation across species. Even the underlying model, when analysed in any detail, will give quite different predictions depending upon many specific facts about the ecological situation. For example, will desertion by the male really lead to possibilities of future matings of which the reproductive benefits will outweigh the possible benefits of caring for existing offspring (Kitcher, 1985: 171)? In fact, Dawkins, sensitive to the variety of

[4] It is also noted that the price that gorillas pay for this glittering prize, apart from the high probability of not winning it, is unusually small testes (Short, 1977). Since their sperm is not forced to compete with that of other males for access to the female ovum, there is no advantage to having a lot of it. On the basis of ratio of testis to body weight, human males are judged to lie in between the gorilla and the chimpanzee, and it is inferred that they are by nature moderately promiscuous.

human mating practices, remarks uncharacteristically that these suggest 'that man's way of life is largely determined by culture rather than genes' (1976: 177). A quarter of a century later, however, few such doubts are entertained by evolutionists. It is to these contemporary versions of human sociobiology that I now turn.

3. Sociobiology Twenty-five Years Later

3.1. Épater les bourgeois

Sociobiologists have always liked to shock. And the picture of the human condition they present is indeed a bleak one. While they usually insist that any possible amelioration of human ills will require the understanding of evolutionary origins, they like to make clear that the origins of these ills are deep and biological. But where twenty-five years ago these pessimistic conclusions tended to be somewhat cautious and speculative, now they are forthright and uncompromising. And nowhere are these shocking conclusions more striking than in the matter of sexuality, as can be discovered by the most casual glance through the biology section of a contemporary bookshop. One does not even need to open the books: on the cover of a book on the human male by British biologist Ben Greenstein, we read:

First and foremost, man is a fertilizer of women.[5] His need to inject genes into a female is so strong that it dominates his life from puberty to death. This need is even stronger than the urge to kill ... It could even be said that production and supply of sperm is his only raison d'être, and his physical power and lust to kill are directed to that end, to ensure that only the best examples of the species are propagated. If he is prevented from transmitting his genes he becomes stressed, ill, and may shut down or go out of control. (1993)

Opinions may, I suppose, differ as to what constitute the 'best examples of the species'. In a slightly more temperate work by the respected evolutionist David Buss (1994; a book which will provide

---

[5] Connoisseurs of sexist language will find this sentence truly breathtaking. If man (generic, surely) is a fertilizer of women, to what species do these beneficiaries of fertilization belong?

my main focus in much of what follows)[6] we also find a depressing message on the dust jacket:

Much of what I discovered about human mating is not nice ... In the ruthless pursuit of sexual goals, for example, men and women derogate their rivals, deceive members of the opposite sex, and even subvert their own mates. (1994)

The emphasis on deception in sexual interactions is a major theme in current biological thought. There is little room for sentimental moralizing in a matter of this importance.

More disturbing still, perhaps, are the following remarks by science journalist and enthusiast for evolutionary psychology, Robert Wright: 'the roots of all evil can be seen in natural selection...The enemy of justice and decency does indeed lie in our genes' (1994: 151). It is no doubt true that if we hadn't evolved we wouldn't do anything nasty. But apart from that rather trivial sense of the 'roots of all evil', it might seem that there are a lot of more immediate sources. But biology, we discover, teaches us that the derelict inner cities, unemployment, and exploitation that we might naively have thought sources of human evil are at most triggers for eliciting our deeply ingrained natural tendencies.

Uniting the popular themes of sex and violence, Buss suggests that men may have an evolved tendency to kill their unfaithful wives under appropriate circumstances (1994: 130–1). If he has anyhow lost control of her reproductive resources he can prevent their being diverted to an evolutionary rival. He may mitigate the great loss in status accruing to a cuckold, and status is important for getting other reproductive opportunities. And—plausibly enough—this will serve as a deterrent to other concurrent or future wives. Wilson and Daly (1992) develop this theme in more detail in terms of their elaboration of the evolved tendency of men to treat women as property.

---

[6] At the time of writing this was the authoritative work on the evolutionary psychology of sex. As I was finishing the manuscript Geoffrey Miller's *The Mating Mind* (2000) appeared, which may come to supersede Buss's book in this dubious role. Miller's book is based on the interesting thesis that the evolutionary explanation of large human brains is a process of runaway sexual selection. I find this thesis quite plausible. It would seem, however, that such an aetiology would make it very unlikely that anything much could be predicted about the behaviour the brain would be liable to emit. The central point about sexual selection is that just about anything can be selected. Disappointingly, then, a quick glance through Miller's book suggests that it recapitulates most of the usual sociobiological claims about sexual behaviour.

They note that some American states until recently treated the killing of a wife discovered in adultery as no crime, and that 'the violent rages of cuckolds constitute an acknowledged risk in all societies, and some sort of diminution of their criminal responsibility is apparently universal' (311). Certainly it is not a pretty picture of our evolutionary heritage.

3.2. The political economy of sex and gender

As I have already remarked, the sociobiology of sex differences has been informed from the outset by an economic metaphor, that of 'parental investment' (Trivers, 1972). The economistic[7] aspects of the field have grown in recent years, and may now fairly be said to dominate it. The central locus of quasi-economic interaction has become the decision to mate. Buss (1994; subsequent page numbers are for this work) entitles two major chapters of his book 'What Women Want' and 'Men Want Something Different'. Evidently we have the classic preconditions for exercise of the fundamental human disposition to—in Adam Smith's famous words—'truck, barter, and trade', and an obvious grounding for the treatment of human relations as a marketplace that has inspired some economists interested in these matters. This perspective naturally invites a consideration of the features men and women will be prepared to pay for in a mate, and his book, Buss notes in the introduction, 'documents the universal preferences that men and women display for particular characteristics in a mate' (8). Put simply, what men want is sex with as many women of as high a quality as possible,[8] and women want to get paid for it. Prostitution, one might say, is the biologically fundamental form of interaction between men and women.

To consider in more detail what women want, their central problem is one of choice among universally eager men. 'Men vary tremendously in the quantity of resources they command—from the

---

[7] In parallel with my use of the term 'scientism' for the view that everything can and should be understood in terms of science (generally quite narrowly conceived), I use the term 'economism' to refer to the application of economic thought beyond its original home in the theory of the production and distribution of commodities. The idea will be central to the discussion in Chapter 6 of the extension of economic thinking to inappropriate areas of human behaviour.

[8] High- and low-quality people are also a central concept for the explicitly economic treatment of sex by Gary Becker (1981/1991); this is discussed further in Chapter 6.

poverty of the street bums to the riches of Trumps and Rockefellers' (Buss, 1994: 23). And, needless to say, this problem is greatly exacerbated by the fact that men will do everything in their power to misrepresent the resources they control in the attempt to dupe women into accepting a less affluent contender than they might otherwise have traded their sexual resources to. In addition, men differ in their willingness to devote their resources to one woman and her children, as to whether, as Buss puts it, they are 'dads' or 'cads'. And again, needless to say, the cads will do everything to convince the gullible woman that they are really dads. (What is not always clear is why there should be any honest dads out there.) The main problem for women, then, is to identify and secure the resources of a Rockefeller dad. Thus women are said to look for various cues in men that signal either the possession, or the likelihood of acquiring, resources.[9] In the former category they prefer, for example, men in suits to those less expensively dressed (101),[10] and also have some preference for men who are older and consequently better heeled (27–8). In predicting future resources, they look for ambition, industry, stability, and intelligence. Women also like a good physical specimen. Apart from the more minimal requirement that their partners be free of open sores and lesions, universally regarded as unattractive (41), women like their men tall. As an extreme illustration of this point, Buss observes that 'when the great basketball player Magic Johnson revealed that he had slept with thousands of women, he inadvertently revealed women's preferences for mates who display physical and athletic prowess' (38). (It might be noted that Magic Johnson did also have some modest resources.) Less anecdotally, but relevant, I suppose, to Magic Johnson, Buss quotes research that is said to show that 'tall men make more money . . . [and] advance more rapidly in their professions' (39). Moreover, they tend to have prettier

---

[9]   Buss reports (24) that women value resources in a mate about twice as highly as men do (the exact number is of course an artefact of his survey design). Given, first, that women in most societies have fewer resources and, second, that women often anticipate dependency on the financial resources of their mates, this is not an observation in obvious need of a deep biological explanation.

[10]   'The same men were photographed wearing either a Burger King uniform with a blue baseball cap and a polo-type shirt or a white dress shirt with a designer tie, a navy blazer, and a Rolex watch' (101). One can't help admiring the attention to detail in the experimental design. 'Based on these photographs women [all women?] state that they are unwilling to date, have sex with, or marry the men in the low-status costumes, but are willing to consider all of these relations with men in high-status garb.'

girlfriends (Buss, 40, citing Ellis, 1992). Apparently this preference for size is not sufficiently explained by the greater resource-acquisition potential of taller men. In addition, women want big men for protection, not a bad idea given the bleak picture of men shortly to unfold.[11]

Finally, in addition to money and size, unless a woman is looking for a fling (something to which I shall return), there is the problem of sorting out the dads from the cads, since the cads, once they have had their way with her, will take off with their resources. What they look for here is signs of love. In all cultures, Buss asserts, women desire love. 'Love is universal' (42). 'To identify precisely what love is', Buss himself has studied 'acts of love' (43). Typical of these are 'talking of marriage, and expressing a desire to have children with the person' (43). The somewhat banal function of these acts of love, when performed by a man, is 'to signal the intention to commit resources to one woman and her children'. Once again, we might worry that the cads are sure to talk the same talk. Indeed in more traditional accounts, this is just what cads are known for.

Men, as I have noted, want something different. The first few subheadings in Buss's chapter on this topic will leave no doubt what this is. They are: 'Youth'; 'Standards of Physical Beauty'; 'Body Shape'; 'Importance of Physical Appearance'; and 'Men's Status and Women's Beauty'. Men, in short, want their women young, cute, and curvy. Evolutionarily, of course, the claim is that men want good breeding stock; and they are prepared to pay for it, even sometimes the high price of (almost) monogamous commitment. That a younger woman will have the potential of producing more children, at least, is not controversial. More surprising, especially to those who have analysed the cultural construction of standards of beauty, is Buss's insistence that these standards are cross-cultural universals. Our ancestors, apparently, needed to assess women for their youth and health. All they had to go on were such features as 'full lips, clear skin, lustrous hair, and good muscle tone . . . a bouncy youthful gait, an animated facial expression, and a high energy level' (53). Somewhat more peculiar is the allegedly universal preference for curves; or, more specifically, a ratio of waist-to-hip measurement of about 70 per cent (57; Singh, 1993). Whatever this supposedly opti-

[11] Though as Philip Kitcher suggested to me (in correspondence), bigger men may also present more of a risk of physical violence to their mates.

mal body shape may show about youth or health, it does, of course provide some useful evidence that the woman is not already pregnant.

This brings us to the one other thing men care about, fidelity. The evolutionary fate worse than death is to invest one's resources in the offspring of another man's genes. Indeed at one point Buss seems to think it appropriate that cuckolders should be required to pay compensation to the victimized husband since this 'reflect[s] an intuitive understanding of human evolutionary psychology: cuckoldry represents the unlawful stealing of another man's resources' (140). Fidelity, however, can be difficult to predict in a potential mate. There is apparently a correlation between premarital and post-marital promiscuity, which suggests that a good cue would be to seek out hitherto chaste women. Oddly, however, while apparently men used to care a lot about this, they do so increasingly less: they still care more in Texas than in California (67), but in Sweden they now care scarcely at all (69). But as I have already noted, evolutionary psychologists are now quite complacent about such minor refutations of their theories. Buss seems happy, in this case, to provide an uncharacteristic cultural explanation of these anomalies.

So far I have considered the generic account of the economic trade between men and women, but with my reference above to the political economy of sex and gender I had rather more in mind. This was made strikingly clear a few years ago when in the course of about a year three long articles on the evolution of human sexual behaviour were published in the prestigious journal *Behavioral and Brain Sciences*. The first presents evidence intended to show that men are attracted to younger women, increasingly younger as they age, and that women are attracted to somewhat older men (Kenrick and Keefe, 1992). The second concerns rape. Specifically, it argues that men have a variety of evolved sexual strategies, and one of these, usually resorted to when others fail, is rape (Thornhill and Thornhill, 1992). The third documents the female preference for men of high status (Pérusse, 1993). Putting the three theses together presents a very simple politics of class and gender: with the acquisition of high status, men have increasing access to women, especially the younger ones they prefer; the lower-status men, having little legitimate access to women, will resort to rape.

These class implications of Buss's story occasionally emerge in striking ways. As mentioned above, the status or quality of both men and women is often crucial to the analysis. For example, 'Men of high status typically insist on more stringent standards for a spouse than most women are able to meet' (50). However, they are 'willing to relax their standards and have sex with a variety of women if the relationship is only short-term and carries no commitment' (50). Occasionally the class markers are more detailed. At one point, for example, Buss describes the predicament of a woman in a singles bar rebuffing the approach of a 'beer-drinking, T-shirted, baseball-capped, stubble-faced truck driver or construction worker who asks her to dance' (144). His angry response, 'What's the matter, bitch, I'm not good enough for you?', is, of course, exactly correct. Buss, I imagine, hopes that she has secured a sufficiently tall protector if she later encounters this low-class specimen in the alley outside the singles bar. Such class stereotypes will strike many readers as quite as disturbing as the gender stereotypes developed throughout the work.

3.3. Methodology

Reading these accounts of male–female relations, one is struck by a mixture of the stereotypic, the outrageous, and the banal. One should not, however, suppose that these are merely the ungrounded speculations of an evolutionist who might better have stuck to ants or seals. I have remarked that evolutionary psychologists do often acknowledge some greater responsibility for presenting empirical data than did earlier sociobiologists, and the claims just cited are constantly buttressed with impressive arrays of empirical data and research. Buss's book synthesizes a thriving and sizeable industry of evolutionary psychological research. Buss reports his own production of thousands of questionnaires on what men and women find attractive in members of the opposite sex, what they take to be significant 'acts of love', and so on. In many, though not all, cases data are offered from a variety of developed and developing countries and from tribal societies, grounding claims of the universality of the phenomena he describes. These are not, it seems, the opinions of an isolated researcher.

Having acknowledged this much, however, closer examination of the empirical data often proves rather disappointing. It will be useful to divide this evidence into categories, which I shall label the absurd, the banal, and the mildly interesting. I shall begin with the absurd.

Perhaps the most glaring example of the absurd is the research, widely cited by evolutionary psychologists, on the hypothesis that men have a mental module the function of which is to measure the waist-to-hip ratio of prospective female sexual partners. The conclusion of this research is that men have a consistent preference for a waist-to-hip ratio of 0.7. The evidence for this curious conclusion is derived first by showing men line drawings of women of various shapes, and asking them which they found most attractive. The presupposition that one could make judgements of this sort on the basis of a line-drawing already incorporates a view of sexual attraction on which it is perhaps politer not to dwell. To buttress this important result, researchers spent painstaking hours poring over back runs of *Playboy* magazine measuring the vital statistics of the models there portrayed with calipers, and again discovered the magic number 0.7 for the waist-to-hip ratio. Since, presumably, the selection of these models reflects men's innate ideals of female pulchritude, the daring hypothesis is further confirmed. Sometimes it is asserted that this shape is also correlated with maximal fertility, though I have not seen, and prefer not to imagine, the research on which this is based. The absurdity of the argument from this evidence to the hypothetical mental module is sufficiently obvious from the fact that evolutionary psychologists much more confidently insist that men are hard-wired to prefer women at the beginning of the fertile stage of the life-cycle. Since hourglass figures are commonest among young, sexually mature women, the results in question would be expected simply as an epiphenomenon of this prior assumed preference. It is, I suppose, possible in principle that men estimate waist-to-hip ratio as a way of detecting young fertile females. But apart from the fact that the research does nothing whatever to support this hypothesis, it seems a highly improbable conjecture. One of the more plausible specialized mental functions of the human brain is the ability to analyse human physiognomy, and it seems unlikely that this undoubted facility would not serve to identify a face as belonging to a young female. Perhaps in the case of androgynous young faces, a glance at the overall shape might be of further assistance in disambiguation. This merely points to the hypothesis that there are a variety of physical cues that have some relevance to the classification of people by age and sex, and that very plausibly people have an ability to integrate a range of cues. A module basing this judgement on a single not entirely reliable gross feature of shape seems otiose.

Equally absurd, though rather less innocuous, is some of the research into the claim that men have a module that directs them, under appropriate circumstances, to rape women. One major source for the claim that rape is a natural male mating strategy derives from experiments done mainly on prison inmates (a questionably representative sample of the population?), referred to in the scientific jargon as 'objective phallometry' (Thornhill and Thornhill, 1992). In these experiments prisoners were made to watch filmed depictions of coercive sex, with instruments attached to their penises that recorded their sexual response to these movies. One variable found relevant to the degree of response was the extent to which the victim enjoyed the incident, a dimension that many experts on this topic would perhaps not consider very relevant to the real experience of rape. Even ignoring problems such as this and assuming that these prisoners were sexually aroused by plausible depictions of rape, the inference that they were disposed to rape has all the persuasive force of the assumption that overweight middle-aged men showing objective signs of excitement in front of their televisions on a Sunday afternoon are disposed to play professional football. (In fairness I should note that Buss, unlike Thornhill and Thornhill, remains agnostic as to whether an evolved strategy of rape has been clearly established [1994: 163].)[12]

Turning from the absurd to the banal, the important point to emphasise in this category is that it consists of claims that most people already believe. The importance of this is that hypotheses that are banal in this sense cannot be taken to illustrate the heuristic usefulness of evolutionary psychology for generating hypotheses. Such hypotheses could just as readily be generated from a casual interview with the person at the next stool in your local bar. In this category are the claims that men prefer somewhat younger female partners and vice versa for women. Of course the fact that such hypotheses are banal doesn't mean that they may not be true, and if they are true it may be a legitimate scientific project to enquire why they are true. I say only 'may be' because there is a subcategory of the banal for which the search for explanation seems wholly redundant. I have in

[12] Miller (2000) argues that rape was unlikely to have been common in the Stone Age. He is required to make this argument, since his central thesis is that sexual behaviour evolved largely in response to female choice among mates, something that would be ineffective if women were commonly subject to coerced sex. Though his arguments here seem plausible enough, I note this fact mainly as an illustration of the ease with which arguments can be made up on both sides of questions about Stone Age life.

mind, for example, Buss's suggestions that evolution has predisposed people of both sexes to prefer partners who are intelligent and kind. The consideration that it might be more amusing to spend a substantial portion of one's life with an intelligent person than with a dullard seems to me to make redundant the speculation that intelligent partners may have been better at distinguishing edible roots or avoiding sabre-toothed cats. But the claims about age preference do seem to provide a sensible occasion for seeking explanation.

The evidence that these preferences are manifestations of innate mental modules is, however, disappointing. The research mentioned above (Kenrick and Keefe, 1992), for example, is based substantially on the analysis of singles advertisements. As with prison inmates, if placers of singles ads form a representative sample of the population, this is something that needs to be demonstrated. But there is a much more fundamental and pervasive problem. These ideas are, as I have said, banal. Most people in most societies think that these kinds of preferences are 'normal' or 'natural'. The media constantly represent couples in which the man is older, often much older. A man of 65 marrying a woman forty years younger excites only mild surprise, and men of that age are sometimes found playing romantic leads in Hollywood movies paired with much younger women. Reversing the gender roles in such scenarios is considered extraordinary. It is reported that typical members of contemporary Western societies watch several hours a day of television, and this points to an obvious way in which such clichés might affect people's assumptions about the normal or the natural. These platitudes might, of course, be platitudes because of imperatives written in our brains by our distant past. But they might also reflect, for example, the fact that men have much greater power in most societies, and the right to youthful partners is one of the exercises of that power. It is not my aim to defend that, or any of an indefinite range of alternative hypotheses one might imagine as to how these social expectations became banal. I want only to point out that the evidence, for example the answers to the questionnaires designed by Buss to elicit the sexual preferences of large numbers of men and women, do nothing to discriminate between these different kinds of explanation. Such raw data are entirely silent on the aetiology of the preferences Buss and others claim to discover. Since in most cases these preferences are clichés—women should be young, narrow-waisted, inexperienced, etc., men should be tall, affluent, sophisticated, perhaps a bit older and more experienced, etc.—it takes little imagination to come up with much simpler explanations than the trials and tribulations of our distant ancestors.

I should perhaps respond at this point to the inevitable tired reaction that I am assuming the Standard Social Sciences Model, a view of the mind as a blank slate on which culture can write as it chooses. I am, of course, assuming nothing of the sort. People certainly have minds of sufficient structural complexity to acquire the dispositions, attitudes, and varieties of behaviour that they in fact acquire. How much structure, and what kind of structure this is, I do not pretend to know. Part of the advantage of my position over that of evolutionary psychologists is just that they do pretend to know. But more important still, there is no reason at all to suppose that a structure that is sufficiently complex to allow human behaviour to be learned will narrowly constrain the kinds of behaviour that can be learned even if, as is by no means uncontroversial, the structure evolved to facilitate fairly specific behaviours that were useful to our Stone Age ancestors. To invoke the computer analogy generally much admired by the scientistically inclined, the fact that the innards of my computer are highly structured doesn't prevent them from carrying out a remarkably diverse set of tasks. And the fact that much of the underlying technology was developed with military applications in mind doesn't entail that my computer is constantly on the verge of planning a nuclear attack, or designing some instrument of mass destruction.

To the obvious objection outlined above, that the evidence adduced in no way favours the hypotheses of evolutionary psychologists over a range of alternative and perhaps intuitively more plausible explanations, one response is to appeal to a range of cross-cultural data. If the same psychological phenomena are found in very diverse cultural contexts, should we not conclude that the phenomena are biologically generated? But this presents problems of its own and although, as I have mentioned, the data that underlie Buss's claims are sometimes collected cross-culturally, very little sensitivity can be discerned to the difficulties of making the relevant cross-cultural comparisons. For example, his insistence that love is a cross-cultural universal is not supported by any discussion of how relevant, and surely quite complex, concepts might be translated unambiguously across cultures. Of course, since he takes love mainly to consist of a disposition to say 'I'd like to marry you and have children with you', the problem may seem to be somewhat mitigated. But in fact this raises another deep difficulty. One of the conclusions that evolutionary psychologists would like to establish is that important anthropological concepts such as 'marriage' have a universal, cross-cultural meaning, a meaning grounded in our evolved psychology.[13] But this is a thoroughly implausible assumption. Anthropologists describe systems of 'marriage' that are monogamous, polygamous, occasionally polyandrous, hypergamous or hypogamous (women marrying up or down in status, though equal status is said to be the commonest case), between people of the same sex, and in some cases as not involving sexual relations at all. And of course there is a wealth of particular rules and expectations surrounding these diverse social institutions. Even within 'Western' culture, the implications of marriage in, say, rural Ireland and Southern California are quite different.[14]

I do not take this diversity to rule out the possibility that these various social institutions may nevertheless reflect the same underlying universal psychology. What I do claim is that evidence about marriage in diverse societies offered in support of such a hypothesis cannot, on pain of blatant question-begging, start with the assumption that these different forms of marriage are fundamentally the same thing. It should finally be added that to the extent that relatively straightforward cross-cultural translation of such concepts is legitimate, it is very likely to be because the cultures concerned have had a good deal of mutual interaction. And of course if this is true, then the value of cross-cultural data is proportionately reduced. And surely the large majority of contemporary cultures do share, to a considerable extent, values shaped by exposure to the same transnational media. It is typical of this kind of work that massive

[13] It is fascinating to note that in their theoretical discussion considered in the previous chapter, Tooby and Cosmides explicitly disavow such universality, since 'purely behavioral categories are seldom able to capture meaningful species-typical uniformity' (1992: 64). However, later in the same volume we read 'Marriage is a cross-culturally ubiquitous feature of human societies' (Wilson and Daly, 1992: 309). This contradiction should be no surprise. As I have noted, taking seriously the complexity of the relations between alleged universal psychology and cultural context would make it difficult or impossible to draw any meaningful conclusions from the behavioural research that occupies evolutionary psychologists when they are not theorizing.

[14] For nuanced discussions of some of these social arrangements around sexuality and gender see, for instance, Ortner and Whitehead (1981). It is also worth noting that for a substantial proportion of the world's population marriages are arranged by families.

collection of data occurs without any real sensitivity to the problems in interpreting the data. Thus the data underlying Buss's story range from the questionable to the ludicrous. As I shall argue in the final part of this chapter, even where the data are clear-cut, there are deep problems in drawing from them the kinds of biological conclusions that Buss wants.

This brings me, finally, to the category of mildly interesting data. Here what I have in mind are empirical results that confirm evolutionary psychological hypotheses that are to some degree surprising (and hence do not belong in the category of the banal). These, as far as I can discover, are thin on the ground. The element of surprise might be in the fact that the hypothesis is confirmed at all, or in the extent of its confirmation. I know of no clear-cut case of the first kind, though probably the best candidate is the research by Leda Cosmides showing that people were much better able to perform simple logical inferences when the subject matter concerned the application of social rules than when it concerned an arbitrary topic. The experiments were a version of the well-known Wason selection task (Wason, 1968). Subjects were given a statement of the form 'If P, then Q', and then shown cards on the visible side of which were statements P, not P, Q, and not Q. They were then asked which cards they would need to turn over to see whether the two sides together constituted a refutation of the statement. Since the statement is only refuted by the conjunction P and not Q, logic requires that the cards P and not Q are turned over. In general subjects proved quite bad at solving this problem where the statement involved, for example, geometrical patterns (e.g. 'If one side of the card has a square then the other has a circle'). Cosmides was able to show that when the statement under test had the form of a social rule, subjects did much better. For a rule such as 'If someone is drinking beer, then they must be over twenty', and shown cards marked 'drinking beer', 'drinking Coke', '25 years old', '16 years old', subjects generally managed to identify the first and last card as loci of possible violations (Tooby and Cosmides, 1992). Cosmides takes this as confirming her hypothesis that there is a mental module serving social cooperation and specifically designed to detect cheats who violate social rules.

I do not want to deny that this is an interesting result, and one that calls for some explanation. The problems, unfortunately, are ultimately just as serious as for the banal cases. Children are constantly exposed to social rules, criticized for violating them, and praised or rewarded for conforming to them. As Cosmides's results confirm, they become very competent at identifying violations of such rules. How could we infer from this the existence of a specialized mental module that produced this result? Explanations have been constructed that assume no such special-purpose module, for example by Patricia Cheng and Keith Holyoak (1989). Tooby and Cosmides (1992) have attempted to show that their data rule out such interpretations, but Elisabeth Lloyd (1999) makes clear that these arguments fail. As Lloyd shows, ultimately Cosmides's argument must fall back on a claim about what must have, or would have been very likely to have, evolved in conditions supposed to have obtained in the Stone Age. But as I have tried to explain in detail, evolutionary theory just can't do this sort of work. Cosmides's research provides an interesting result for cognitive psychology, but does nothing to settle questions about the extent of innate structure in the brain.

The most striking quantitative surprise claimed to the credit of evolutionary psychology is the data from Daly and Wilson on the discrepancy in the amount of violence to children perpetrated by step-parents and biological parents. No one would be surprised to learn that there was some such discrepancy: most of us are familiar with the sad plight of Cinderella, and the idea that her situation is a not uncommon one perhaps belongs in the category of the banal. Daly and Wilson (1988), however, showed that using actual homicide as an index of violence against stepchildren or adopted children, the occurrence of this was many times that for biological children . There is no doubt that there are social factors that would predict some of this difference. Perhaps there are biological grounds for the prevalence of the view that 'blood is thicker than water', but it is at any rate a view widely held to be true. And no doubt it is widely assumed that there is a natural human goal of producing and raising one's biological offspring. Equally true and important is the fact that for every child murdered by its step-parents there are hundreds or thousands brought up by step-parents who provide just as much care and love as most biological parents. So we have a rare but horrible breakdown of the norm of parental care that occurs much more frequently for non-biological than for biological parents. We have some obvious cultural factors that go some way to account for this discrepancy, but perhaps not far enough. Any parent will testify that it is easy enough to see why, if one did not feel affection towards children, one might well murder them. So perhaps there is a biological disposition to feel affection for one's own offspring that helps to prevent this unfortunate outcome. On the other hand, it must be reiterated that in the vast majority of cases this biological deterrent is redundant, as shown by all the non-biological parents who show no disposition to murder their children. It is entirely unclear what inference should be drawn about the nature and action of whatever innate disposition one may have to care about the genetic origins of one's children.

A final point is worth mentioning. Recent research, no doubt disturbing to many men, has suggested that somewhere in the region of 15 per cent of children were not in fact fathered by the men who take themselves to be the biological father. It would, no doubt, be a persuasive bit of evolutionary psychological evidence if these men were found much more likely to commit violence on their children. But in the absence of such evidence, I conjecture that such a correlation would hold only to the extent that these men knew or suspected that they were not the biological fathers. If that is the case, then the phenomena under consideration work through conscious cognition.[15] And that, in turn, suggests that they should be susceptible to the influence of social norms. This is not, of course, an argument against there being a biological component to what is, certainly, an evolutionarily fundamental social relationship. I do want to insist, however, that the evidence under consideration licenses no compelling conclusions about the innate structure of the mind.

4. Further Reflections on the Poverty of Evolutionary Psychological Inference

In this section I shall further explore the difficulties in the attempt to infer from psychological phenomena to evolved functional components of the mind. First, however, I would like to mention another strategy somewhat notoriously connected with sociobiological thinking, the comparison of human behaviour with that of the behaviour of other species. Sociobiologists have often been accused, and often with justice, of supporting their arguments by appeal to any convenient non-human species that happened to behave in an apparently analogous way. Thus, for example, scorpionflies and

---

[15] If my hunch is wrong, of course, that would indeed provide genuinely persuasive evidence for an unconscious, perhaps even innate, mental mechanism.

ducks have figured largely in discussion of the alleged biological roots of rape.[16] In criticizing such strategies it has been noted first, that the examples were often arbitrarily selected; and second, that only in the crudest analogical sense could, for example, the behaviour of copulating flies be related to that of human rapists. It is fair to say that contemporary evolutionary psychologists depend less heavily on this strategy than their predecessors, in part, of course, because they claim much more data derived directly from the study of humans. However, animal analogies still play an important rhetorical role in this work, and sometimes seem all the more bizarre for their lesser frequency.

To take a few examples from Buss: 'Women, like weaverbirds, prefer men with desirable nests' (7); or, 'Like the male roadrunner offering up his kill, men offer women resources as a primary method of attraction' (100); and 'humans' ways of solving the adaptive problem of keeping a mate are strikingly similar to insects' (124). The latter include such methods as physically carrying the female off to some place less frequented by competitors or, which sounds to me distinctly unlikely as a human strategy, shedding their broken-off genitalia after copulation to seal off the reproductive opening of the female. I shall not dwell on this issue because, as I have noted, it does not play an obviously central role in the kinds of arguments I am considering. No doubt part of the function of this constant ornamentation of the text with these more or less fanciful parallels is to remind the reader that the author is, after all, doing no more than taking seriously the fact that we are ultimately just animals. Whether anything much follows about any specific kind of animal merely from the fact that it is, ultimately, just an animal is another matter.[17]

The empirical detail characteristic of contemporary sociobiology raises a further difficulty that I want to stress. A common objection to earlier variants of sociobiology was that their accounts of human behaviour were massively simplistic. Modern evolutionary psychology has partially responded to this objection, and provided

---

[16] See Fausto-Sterling (1985) for trenchant criticism.

[17] I do not, of course, mean to deny that there is a role for comparative phylogenetic studies in establishing the adaptive nature of traits. But this role requires evidence that a trait is homologous between related species. Trawling through the animal world for analogous traits, as in the examples in the text, has no such value. As Jonathan Kaplan (in correspondence) has emphasized to me, the unusual lack of close relatives of humans makes the legitimate strategy largely unfeasible.

accounts that allow for more complex and varied behavioural strategies. But in doing so it has exposed even more clearly than before the difficulty, emphasized many years ago by critics such as Gould and Lewontin (1979), that the theory is almost infinitely malleable and consequently empirically empty. To consider one example, early emphasis on the evolution of pair-bonding as well as on a male tendency to promiscuity seemed to some not only simplistic, but also as verging on the inconsistent. The obvious difficulty derives from the tautological, but still sometimes neglected, observation that the total number of matings by males and females is identical. Given that there is an approximately equal number of heterosexual males and females, the average number of matings per male and female will also be the same. (It is true that the proportion of males does tend to decline with age, but not to an extent that is relevant to the general point.) And, since at least the Kinsey Reports, it has been scientifically well established that humans, in both sexes, are variably but moderately promiscuous animals.

The more empirical turn in contemporary evolutionary psychology has taken account of these facts. In place of earlier monolithic theories of the sexual predilections of men and women they have suggested a repertoire of evolved sexual strategies. (The suggestion that rape is an evolved alternative sexual strategy for otherwise unsuccessful men is an example of this manœuvre.) Typically, the idea is that in addition to psychological mechanisms designed to promote pair-bonding, humans have alternative strategies for engaging, under appropriate conditions, in casual sexual liaisons. Within the evolutionary framework it is not difficult to see why men should be said to have evolved this strategy, either before or after engaging in pair-bonding. However, recalling again the tautology mentioned in the previous paragraph, some account is required of why women might cooperate. In fact, without some chance of finding amenable women, there is no evolutionary explanation of the male tendency to casual sex: looking around for opportunities for casual sex when none are to be found is, presumably, a mere waste of resources and should be penalized by evolution. Thus a major growth industry in evolutionary psychology is the provision of explanation for female proclivities towards casual sexual encounters.

Unsurprisingly, the main thrust of such explanations is once again economic. Apart from prostitution in the strict sense, women are perceived as providing themselves with insurance against the provisioning inadequacies of their principle mate. Buss spells out the prehistoric scenario:

Imagine a food shortage hitting an ancestral tribe thousands of years ago. Game is scarce. The first frost has settled ominously. Bushes no longer yield berries. A lucky hunter takes down a deer. A woman watches him return from the hunt, hunger pangs gnawing. She makes him an offer for a portion of the prized meat. Sex for resources, or resources for sex—the two have been exchanged in millions of transactions over the millennia of human existence. (1994: 86)

In slight twists on this simple economic tale, women are said to be providing insurance (their mate may lose status or command of resources or, for that matter, die, so they are establishing connections with possible replacements) or to be setting up a network of provisioners.

A different kind of story suggests that women may perceive that the man who is the best provider that they can secure may not have the best genes they can attract. Thus they might attempt to get their genes from a different source.[18] In support of this hypothesis, empirical evidence is said to show that married women usually have lovers of higher social status than their husbands; that they arrange trysts with their lovers disproportionately while they are ovulating; and that they have more orgasms with their lovers than with their husbands. (Female orgasm is now said to cause more sperm to be retained in the reproductive tract.) Husbands, incidentally, are said to respond by ejaculating higher numbers of sperm when their wives have been out of their sight, thus attempting to swamp the contributions of their suspected competitors.

A rather more bizarre explanation of female promiscuity might be called the self-appraisal theory. In the context of the general economic metaphor, it is important for a woman, especially, to have an accurate idea of her market value. By engaging in a series of casual sexual encounters she can, on this account, 'obtain valuable information about the quality of the men she can potentially attract' (Buss, 1994: 89). She thus avoids the twin dangers of selling herself short, and of holding out for more than she can command. (The fact that,

---

[18] This ingenious, if Macchiavellian, strategy, is attributed to various bird species. See Wilson and Daly (1992: 292–7) for an account of avian sexual shenanigans in the swallow and dunnock, and numerous further references.

according to another part of sociobiological theory, she will, as a consequence of her value-appraisal exercise, also reduce her value by becoming more sexually experienced, creates the sort of problem beloved of mathematical economists.)

These various accounts illustrate plainly the ease with which evolutionary stories can be constructed. Early sociobiological intuitions about female monogamy are readily superseded by a host of complicating adaptive considerations. With sufficient ingenuity multiple possible evolutionary benefits can be imagined for almost any form of behaviour. And this, of course, shows only that such stories should be treated with great scepticism. This scepticism should be amplified when, as in the present case, a whole series of alternative stories are offered for the same supposedly evolved behaviour.

But perhaps an even more important point is the way in which the attempt to accommodate the empirical variability of human behaviour leads to the introduction of ever more flexible, and arguably ad hoc, auxiliary assumptions. If a behaviour is thought to be more or less universal across cultures it is because it evolved. If there is an exception (such as the lack of concern by men about premarital female promiscuity in Scandinavia) it is because there is sensitivity to cultural influences. As Buss puts it, 'some preference mechanisms are highly sensitive to cultural, ecological, or mating conditions, while others transcend these differences in context' (1994: 254). It is, of course, equally possible that the social conditions that encourage some of these preferences are currently less variable than those that support others. At any rate, it is clear that once these strategies are admitted to be subject to cultural influence, any amount of variability will be fully explicable within the sociobiological paradigm. And as is a familiar truism in the philosophy of science, a theory that can explain anything explains nothing.

The ease with which evolutionary psychologists can accommodate data is strikingly illustrated in a paper by Bruce Ellis commenting on the fact that in questionnaires women, contrary to evolutionary prediction, claimed to attach little importance to either dominance or social status. Ellis offers four possible explanations: they may mistakenly have supposed that the men were disposed to dominate them rather than other men; they may be reluctant to admit that they prefer such men; they may prefer such men but be unconscious of the preference; or their assumed reference class may only include high-status men, among whom details of status will not be important (Ellis, 1992: 282). Perhaps so. But philosophers of science have long seen such multiplication of auxiliary hypotheses, hypotheses introduced solely to account for a failure of match between theory and actual experience, as the main symptom of a theory in decay. In the terminology of Imre Lakatos (1978), these are the signs of a degenerating research programme—if, indeed, such a judgement does not imply more antecedent progressiveness than is evident.

Let me conclude this section with a brief comment on the great difference between the context in which, according to evolutionary psychologists, the psychology of sex evolved and more modern conditions. Even if our hypothetical cavemen ancestors selected mates solely on the basis of their reproductive potential, things have got a bit more complicated. So-called trophy wives would not, perhaps, be accounted trophies if there were not some recognized virtue to mere youthful good looks; but a trophy wife seriously deficient in intelligence, charm, good manners, etc. would, I suppose, be as often an embarrassment as a prize. Prudent mate-selection, that is to say, involves a wide range of factors, many of which have nothing whatever to do with purely physical attractiveness. Although evolutionary psychologists do mention a range of such factors, the attempts to explain the importance, for example, of intelligence or kindness (Buss, 1994: 34–5, 45) in terms of effects on fertility are both implausible and redundant.

Of equal importance is the fact that mate-selection, in the sense of selection of a long-term partner for the bearing and rearing of children, is hardly the sole context in which modern humans make judgements about the attractiveness of other people. Whether or not this was true of our less sophisticated ancestors, contemporary humans are interested at different times in a variety of different kinds of relationships with members of the opposite sex (or, in many cases, the same sex; though how this relates to the present issue is obviously problematic). They may seek friendship, casual sex, a brief romance, lifelong companionship, a co-parent for their children (existing or yet to be born), a status symbol, a domestic drudge, and so on. Presumably the relevance of prehistoric whisperings concerning reproductive potential will vary considerably from one to another of these cases.

These considerations emphasize why, regardless of evolved psychology, we should be in no way surprised that sexual behaviour is highly varied, and hence reinforce the impossibility of inferring the evolved psychology from behavioural data. This can be best seen in terms of a very general worry about the allegedly massively modular mind. However modular the mind may be, the output of such modules must somehow be integrated into some broader process in which whole human beings come to make decisions, and must be capable of weighing modular outputs differently according to different ends to which the decision-making process may at any time be directed. There must be some part of the mind in which it is possible to decide whether to pursue a potential mate or forage for carrion in the nearest fast-food outlet. Suppose, for the sake of argument, that there is indeed a mechanism in the human brain that disposes men to select very young women or girls as ideal mates. Given that this atavistic mechanism provides only one of a range of inputs into actual processes of mate-selection, and given that mate-selection, in the sense assumed by evolutionists, is only one of a range of kinds of behaviour in which this hypothetical machinery might figure, it is not at all clear that identifying such machinery will tell us anything much about the behaviour or even behavioural dispositions of modern humans. At the most, we might learn something about psychopathology: the maladapted mind, the mind unable to function in the conditions in which it finds itself, is perhaps a mind constantly and uncontrollably driven by atavistic urges from its evolutionary past. The healthy mind, the mind that despite its Stone Age origins functions effectively in the complex context of modern life, is another matter.

In summary, then, the evolutionary psychology of sex and gender offers us mainly simplifications and banalities about human behaviour with little convincing illumination of how they came to be banal. It offers us no account of the great differences in behaviour across cultures, which is exactly what we might want to know if we were interested in exercising any measure of control over the changes in these phenomena. It offers no account of why different people develop such diverse sexual proclivities (notoriously, it has nothing but the most absurd evolutionary fantasies to offer in explanation of homosexuality). And it offers no account of how the complex motivations underlying sexual behaviour interact with the pursuit of the many other goals that inform the lives of most humans. In fact it offers us nothing, unless perhaps a spurious sense of the immutability of the behaviours that happen to characterize our own contemporary societies. This is scarcely the revolution in our understanding of human behaviour so enthusiastically advertised by the exponents and camp-followers of evolutionary psychology.

I have looked at just one area of evolutionary psychological speculation, though perhaps the most active one. But most of the difficulties discussed would apply in very similar ways to other areas to which such methods might be directed. Cultural variability and individual variability can be found in most interesting domains of human behaviour, and all areas of behaviour come about as the upshot of complex negotiations between motivations of various kinds. Evolutionary psychology has provided no resources for dealing with problems of these kinds. In relation to the illumination of the real complexities of human nature, the programme may be declared bankrupt.

4

The Charms and Consequences of Evolutionary Psychology

1. Introduction

I now want to turn to some broader issues that arise from the fore-going discussion of evolutionary psychology and its epistemological worthlessness. First, I want to consider the question why, if I am right in claiming that the programme has such severe methodological weaknesses, it has nevertheless attracted such a wide degree of sup-port and public interest. This leads me to introduce some general ideas about the nature of science which I take to be both widely held and deeply mistaken. Some of the appeal of evolutionary psycholo-gy, I shall argue, is based on such misconceptions of the nature of sci-ence. But second, in addition to such philosophical error, it is natural to look at the attractiveness of particular scientific projects in terms of the professional aspirations of scientists. I shall consider some aspects of evolutionary psychology, aspects it shares with several other scientific projects I consider equally misguided, that might pre-sent such professional attractions. And third, one very understand-able basis of the appeal of evolutionary psychology is that if it can give us important insights into human nature, it is natural to con-clude that it has implications for how people should live or how soci-eties should be organized. Curiously, perhaps, these implications have been much more insisted upon by critics than by practitioners, and the latter have tended to deny that there are any such implica-tions. This is the final set of issues that I shall explore in the present chapter.

2. The Epistemological Charms of Evolutionary Psychology

2.1. The unity of science

It is harmless, and perhaps even beneficial, to suppose that every-thing we want to know about the world should be addressed in a scientific spirit, if that means only that we should carry out research with sensitivity to such evidence as is available, theoretical coherence, and no doubt other equally bland intellectual virtues. But generally when scientists advocate a new approach to some important area of inquiry they will not only advertise their approach as more truly scientific than those they aim to supersede, but they will have some implicit or explicit assumptions about what it is to be scientific. But as I have argued at length elsewhere (Dupré, 1993a), beyond the attainment of a maximal number of the bland intellectual virtues, there is nothing that it is to be scientific, no common essence to the diverse and variously productive projects generally collected under the heading of science. The contrary assumption is not only mistaken, but can be seen to justify bad, and even sometimes harmful, science. Such is the case for evolutionary psychology, or so I maintain.

Tooby and Cosmides begin their long introduction to *The Adapted Mind* (Barkow, Cosmides, and Tooby, 1992) with a section entitled 'The Unity of Science'. In philosophical discourse the doctrine of the unity of science has been associated with various stronger or weaker versions of reductionism. But Tooby and Cosmides appear to mean something much weaker than this. At first it appears that all they propose is that the sciences should be interconnected and mutually coherent. As long as not too much is smuggled in under the concept of interconnection, these are exceedingly weak demands. It is perhaps plausible that they were violated by ontological dualisms such as that between celestial and terrestrial mechanics before the sixteenth century, between mind and matter for Cartesian dualists, or by those who postulated an immaterial vital fluid that imbued dead matter with life. So perhaps what Tooby and Cosmides have in mind is a thesis of ontological monism or ontological unity: the only stuff of which anything is made is physical stuff: there are no Cartesian minds, souls, or ghosts. This is a serious philosophical claim, and one that has only recently become widely accepted. But it has little connection with reductionism or indeed any other doctrine about how science should proceed. The idea that there is just one kind of stuff (allowing that the many bits and pieces physicists now distinguish are, in some non-trivial sense, of the same kind) hardly entails that phenomena of all kinds can be fully understood merely as consequences of the properties of that bare stuff.

But Tooby and Cosmides do seem to want to extract a substantive view of how science should be done solely from the assumption of ontological unity. This view, though weaker than full-blown reductionism, is much stronger than merely the assertion of coherence and interconnectedness, and might be called the thesis (I would prefer to say myth) of the centrality of the endogenous. This is the thesis that we should always favour explanations in terms of the intrinsic, structural properties of things over explanations that appeal to the influence of context or environment. Though falling short of explicit reductionism, it is easy to see that this idea shares the same aetiology, namely the idea that ontological unity implies some kind of explanatory preference for the constituents of things over their contexts. As I shall argue in Chapter 7, I do believe that when ontological unity is combined with a further powerful thesis, the causal closure or completeness of physics, something like this becomes compelling. But for now I note only that without some such supplementation, ontological unity has no consequences at all for strategies of explanation. Ontological unity does not imply that the causal properties or relations of complex entities are simply consequent on their structural composition unless, that is, one assumes full-blooded reductionism.

One example of this thesis of the centrality of the endogenous in action is the exaggeration of the role of the gene in development, and especially the development of the brain which, except from a myth-eaten perspective, is quite obviously strongly dependent on the environment. Evolutionary psychology can be seen as a further expression of the same myth. Despite lip service to the importance of interaction with a variable environment, strategies of investigation that start with the influence of the environment are seen as worthless and the only properly scientific way forward is investigation of the intrinsic tendency of the organism to emit behaviour of various sorts. The environment ends up as little more than a trigger that determines a selection from among the range of internally generated behaviours.

Everyone agrees that actual human behaviour has both internal and external causes, so why should one favour one kind over the other? The answer, as I have been suggesting, is a covert assumption of a kind of reductionism. Causes may come from inside and outside a system, but order comes only from inside. Order derives from complex, machine-like physical structures exploiting the only ultimate laws there are, the laws of physics. And ordered phenomena are where we must look for explanation and understanding. Thus in explaining their preference for evolutionary psychology over what they call the Standard Social Sciences Model, Tooby and Cosmides remark: 'the SSSM induces researchers to study complexly chaotic and unordered phenomena, and misdirects study away from areas where rich and principled phenomena are to be found' (1992: 23). Or: 'human culture is richly variable because it is generated by an incredibly intricate, contingent set of functional programs' (24). Treating a set of phenomena (culture, or structured patterns of human behaviour) in terms of properties of cultural objects themselves (social institutions, processes of education, shared beliefs, and so on) reveals only chaos. The chaos reveals itself as ordered only when we see it as generated from the orderly properties of the structural constituents (human minds). And this idea, the transformation of disorder into coherent patterns by retreating to a lower structural level, is the standard theme in more robustly reductionistic accounts of the unity of science.

Any superficial plausibility there may be in the idea that the orderly must be traced to the internal workings of things is easily dispelled. How is the order found in human brains to be explained? Let us allow for the sake of argument that it is by appeal to the storing of information in human genes. But this order, at least, is to be explained by appeal to evolution by natural selection, which is of course a higher-level process external to the individual organism. If processes at the organismic level can figure in a fundamental explanation of the behaviour of parts of organisms, then there can be nothing conceptually amiss with the thesis upon which Tooby and Cosmides pour such scorn, that the dispositions of human minds are to a substantial extent the product of social forces. Certainly this no more violates any conception of the unity of science than does their own appeal to natural selection. And no hint of incoherence or inconsistency is in the offing between these various and perhaps interacting processes. So I conclude that Tooby and Cosmides's rather vague appeals to the unity of science are wholly irrelevant to the plausibility of evolutionary psychology.[1]

---

[1] A more general problem arises once the iterative nature of the strategy of seeking order in terms of the orderly behaviour of constituent parts is addressed. When, as in the present case, the constituents are themselves structurally complex, why should we expect any more superficial order at that level than at the higher level? Either structurally complex entities can behave in orderly ways anyway, in which case no general argument can be offered against looking for order in culture itself, or they cannot, in which case there is no

The distinguished philosopher Daniel Dennett, already cited as an enthusiastic supporter of evolutionary psychology, has attempted to defend evolutionary psychology by appeal to similar considerations, in this case by means of a forthright endorsement of reductionism. The defence, however, is no more compelling than that of Tooby and Cosmides. Dennett begins by distinguishing preposterous reductionism, the idea that all of science could ultimately be replaced by physics, from bland reductionism that unifies sciences only in the most minimal sense of requiring that they avoid contradicting one another (1995: 81). Thus animals, since they are, among other things, physical objects, will obey the law of gravity. This is slightly reformulated into a contrast between greedy reductionism (bad) and good reductionism. Good reductionism is 'simply the commitment to non-question-begging science without any cheating by embracing mysteries or miracles at the outset' (82). What remains something of a mystery is what Dennett means by a 'mystery'. If it means something along the lines of its technical sense in Catholic theology (the Trinity, or transubstantiation, for instance) then few scientists would demur. If it means that everything assumed in a scientific explanation must itself be fully explained, then science would be wholly impossible. It is true that the evolution of culture or the ways that it acts on individuals are not matters that are fully understood. But if that makes culture a mystery then the mental modules of evolutionary psychology are every bit as mysterious. So this bland reductionism is entirely neutral as to the debates currently under review.

There is another, quite different sense of reductionism that Dennett evidently has in mind, which is summed up by another quote a few lines after the last. 'Darwin's dangerous idea [evolution by natural selection]', he writes, 'is reductionism incarnate, promising to unite and explain almost everything in one magnificent vision.' This is an example of something rather different, if related: what I prefer to call scientific imperialism. By this I mean the tendency to push a good scientific idea far beyond the domain in which it was originally introduced, and often far beyond the domain in which it can provide much illumination. I shall return to this topic later in this chapter. It is an intellectual disorder that is the subject of criticism at several points in this book.

advantage in stopping with the individual's mental modules or, indeed, anywhere short of particle physics.

I conclude that whereas reductionism clearly has a role in explaining the attractions of evolutionary psychology, the appeal to any variety of reductionism that is at all defensible has little bearing on the argument for evolutionary psychology. No doubt the suggestion that the human sciences should depend more heavily on biology than has previously been allowed will appeal to a reductionist sensibility. And no doubt it does us good to acknowledge piously that we are animals evolved from animals, and what we say about ourselves should be consistent with that fact. But not even the most rabid evolutionist denies that behaviour varies in significant ways according to social context, and even the most die-hard culturalist agrees that we are organisms of a specific if highly variable kind, and that accounts of behaviour must be constrained to some degree by the kind of organism we are. So the real questions still concern the ways in which our biological nature interacts with countless aspects of our environment to produce behaviour. And these are not questions that can be resolved by abstract commitments to any reductionist thesis.

2.2. The genetic fallacy and the fear of contingency

Evolutionary psychologists like to portray themselves as the standard-bearers of Darwinism. They, but not their opponents, are truly taking Darwin seriously. The assumption is that if we fully accept that we indeed evolved from earlier non-human animals, and by fundamentally the same processes as other animals, then this must provide us with the proper key for disclosing the nature of human nature. But so far the argument has little merit. It is true that Darwinism has a vital negative role with respect to human nature. From its inception to the present day, its most virulent opposition has come from those who think of humans as divinely created. Creationism, at least of the fundamentalist varieties that oppose Darwinism, does provide its adherents with a clear method both for understanding human nature and for discovering how it is ethically right for humans to live and behave, namely, careful study of Holy Scriptures. Darwinism undercuts the grounds of the creationist perspective, and few who have accepted the truth of evolution think that scriptures can suffice to play such a role.

But here, ironically, evolutionary psychologists have retained assumptions that should have been rejected with special creationism. For contrary to both these schools, there is no reason why an account of origin should play anything like the central role in understanding an entity or phenomenon that it plays for fundamentalist creationists. This is most obviously true for social phenomena. We (or some of us) know quite a lot, I suppose, about the history of the United States of America. But it seems entirely fanciful to suppose that, for instance, the decay of racially segregated inner cities could have been predicted in, say, 1776. Of course, when we explain such a phenomenon we appeal to historical factors such as slavery. But historical explanations are complex and partial, involving the accumulation of many different factors at different times. They are also shot through with contingency. The trajectory cannot be extrapolated into the distant future. Although historical factors leave their mark, what is currently happening in American inner cities depends on the current state of the economy, policing practices, housing policies, and much else that is entirely contemporary. Perhaps it also depends on evolved dispositions to xenophobia that can be traced to the Stone Age; but these, if such there be, will do nothing to explain the difference in the frequency of violent crimes between Detroit and Paris, or London and Cape Town.[2]

I suspect that part of the problem here is a very fundamental ambivalence common to many evolutionists. On the one hand evolution is universally seen as a scientific replacement for an earlier mystical view of the origins of life, a perspective with which I have much sympathy. On the other hand, it is quite obviously a historical subject, one that traces the particular vicissitudes of an exceedingly complex process on a particular seemingly insignificant celestial body. It is still common to think of science as showing how things had to happen the way they did by discovering the inexorable laws of nature that made them happen. But history, surely, is not like this.[3] Laws may perhaps play a role in connecting specific events, but the view that the whole sequence of historical events is inexorably determined is a thesis in metaphysics not the theory of history. History itself is an indissoluble mixture of processes that seem more or less inevitable once under way, and entirely contingent events.[4] Much

---

[2] For a useful discussion of attempts that have, nevertheless, been made to explain violence and criminality genetically, see Kaplan (2000: ch. 5).

[3] Or at least not after the first second or two.

[4] The sense of contingent I have in mind is perfectly captured by Aristotle's example of meeting someone *by chance* in the marketplace. However fully the factors that make each of us go to the marketplace may necessitate that we go there when we do, the fact that

misguided evolutionary thought can be seen as the futile attempt to make history necessitate and thereby to make a thoroughly historical study conform to the idea of science as the discourse of natural necessity.

Arguments that Darwinism must provide the key to understanding human nature are naturally often more subtle than a vague gesture towards genesis, however. Such arguments generally pick on an aspect of the biological, and a fortiori human, world that can only be explained by appeal to evolution by natural selection, and claim that this aspect is what is fundamental to understanding biological organisms in general and ourselves in particular. Such allegedly fundamental perspectives are indicated by such concepts as adaptation, function, or design. Although these are importantly different concepts, all of which have been subject to detailed analysis by philosophers, they all point to the same central idea. Organisms are highly complex structures, and aspects of their structures provide them with remarkable capacities to deal effectively with their environments. A paradigm is the eye. The eye, together with the parts of the brain that process information from the eye, is an exquisitely organized device for gathering information about the environment that surrounds an organism, and thereby for facilitating appropriate responses to that environment. It is impossible to avoid the conclusion that the function of the eye is to gain this information about the surrounding environment or, more simply, to see. Relating this more explicitly to evolution, it is said that the eye is an adaptation for seeing. Philosophical analyses of the concept of adaptation often explicitly include the assumption that an adaptation is a feature of an organism that exists because of the benefits it provided for ancestors.[5] Surely the eye (whether of a human or an octopus) evolved by natural selection, and the benefit that decreasingly rudimentary eyes provided for our ancestors was that they provided increasingly sophisticated knowledge of their surroundings. So the eye is an adaptation for seeing.

Somewhat more problematically, it is sometimes said that evolution *designs* features of organisms. This is a central idea in Daniel Dennett's extended hymn of praise for Darwin's theory (1995).

we are both there at the same time is not necessitated in this kind of way. Similarly many historical events lie outside the scope of whatever laws of history there may be.

[5] Many of the most influential discussions are collected in Buller (1999).

Clearly this should be seen only as a metaphor, as designing things is something done literally only by intelligent beings. An important feature of the metaphor is the continuity it provides with the earlier idea that organisms were quite literally designed by a supreme being. It is, I think, very important to insist on the difference between adaptation and design. The transition from a concept of design to a concept of adaptation was after all the great achievement for which Darwin is justly famous, and one might fairly argue that the constant confusion of the two ideas that runs through Dennett's book just cited marks him as a thoroughly pre-Darwinian thinker.

I said the eye was an adaptation for seeing. But it was not designed for seeing, since no one went to the trouble of designing it. One very important consequence of this is that it makes problematic a certain kind of atomism that is often assumed in analysing design. When designing the air-conditioning system of a car, say, one takes the rest of the vehicle as fixed and works out how best to cool its interior. This makes plausible the project that might be carried out by a rival car manufacturer of reverse engineering, trying to work out why the first manufacturer has done things the way they have by assuming that they were intelligently, perhaps even optimally, addressing the problem of cooling the interior of an automobile. Dennett considers this strategy, trying 'to figure out what reason, if any, "Mother Nature" . . . "discerned" or "discriminated" for doing things one way rather than another', to be an 'extremely fruitful and, in fact, unavoidable' one for dealing with organisms. The attribution of rationality to 'Mother Nature' is of course an ingenious way of converting history (natural history) into necessity. For as theologians have long understood, perfect rationality constrains the agent to only one possible action, which is to say that it necessitates an action.

Any design process takes place under constraints. The air-conditioning designer must start with the kinds of refrigeration units available or readily manufacturable, must find somewhere to mount it that is not already occupied by other essential components, and must generally take the shape, airflow dynamics, and so on of the vehicle as given. But the constraints on 'design' facing Mother Nature are different in kind and in degree. Mother Nature does not, for instance, start off with a sightless human and work out the best way of equipping this creature with sight. It may be, as Richard Dawkins speculates, that the eye developed over aeons of time from a patch of light-sensitive cells somewhere on the surface of the body (1986:

77ff.). But the ancient creatures with these patches of light-sensitive cells were nothing remotely like us. (Which is just as well as such a patch would probably not do us much good.) Each of the many stages between this ancient proto-eye and a modern human eye had to serve the particular kind of creature that was its happy possessor. Moreover the transition from, say, a creature with a patch of light-sensitive cells to a creature with a concave indentation filled with light-sensitive cells must have been constrained by the developmental and genetic possibilities of those particular organisms. And, as is well known, making genetic changes to an organism doesn't typically make one local change, but often generates many changes that ramify through the developmental sequences of the organism.

None of this is supposed to be an argument that natural selection cannot produce adaptations. Natural selection provides by far the best, and perhaps the only, account we have capable of explaining the kind of functionality we find in physiology. But the process is a historical one, constrained at every point by the historical contingencies of the moment. This is the transformation that Darwin brought about from earlier theories of design, and it is a transformation which, curiously enough, thinkers such as Dennett seem not to have grasped. The point is perhaps clearer if we consider another class of historically generated but more or less functional entities, political institutions. Take, for instance, the British House of Lords. It is clear enough what it does, namely debate and sometimes obstruct legislation. This is arguably a useful function in a political system, but no imaginable exercise in reverse engineering could predict the structure of the institution serving this function. The reasons why this role should be carried out by the descendants of the mistresses of ancient hereditary rulers, or the leading expositors of a largely defunct religious ideology, could not possibly be discovered by reflections on engineering. The explanation, of course, must be sought in terms of the historical contingencies at the time at which these particular features of the institution became established. And, as with biological organisms, once established it is extremely difficult to change, piecemeal, features of the system. Fortunately, political institutions have an element of real design as well as historical constraint, so it is possible, as recently demonstrated, to eliminate intentionally the most egregiously absurd features of an institution such as the House of Lords. No doubt with the passage of time Mother Nature can accomplish the same for organisms; but, to reiterate a central point, by the time she pulls off this trick it will be for our distant descendants, creatures no doubt very different from ourselves.

The fact that organisms originated from a long, complex, and constantly constrained historical process has momentous consequences. To the extent that it is possible to understand them by reflecting on their origins it must be in terms of this history, and this must take seriously the details of history over aeons. A project as simplistic as reverse engineering has no chance of pulling off this trick. Whereas recent history (centuries or perhaps millennia) may be accessible enough to provide us with some insight into differences between contemporary humans, the ever more distant history relevant to the common biology of humans is perhaps forever too vaguely known to provide a useful source of such insight. It certainly seems a more promising project, though of course a daunting one, to investigate more directly the product of natural selection, ourselves, than to attempt to infer the product from the process.

It is worth adding, finally, that not only human evolution, but also individual human development, are thoroughly historical processes. As has been emphasized in earlier chapters, the relation between genotype and phenotype is not one between blueprint and product, but one in which the genome is but one among a range of interacting factors that combine to produce a complex and in some respects unique product. These two importantly historical processes are involved in the production of individual humans, and the contingency of these historical processes lies at the heart of the inadequacy of the kind of reductive science that evolutionary psychology exemplifies.

How might we resolve the tension between science and history? Only by fully accepting the disunity of science, or the diversity of the sciences. As a historical study evolutionary theory will be a very different activity from sciences such as mechanics or chemistry that deal with timeless properties of things.[6] Once we give up the idea that there is something common to all sciences that uniquely equips them for the production of knowledge, and the corollary that all sciences

---

[6] It should not be forgotten that chemical species also evolve in a certain sense. It does appear that the early stages of this process are fully determined by intrinsic properties of matter, so that the appearance of atoms and simple molecules was necessary given the history of our universe. The appearance of complex biochemicals such as DNA, on the other hand, appears a much more contingent matter.

function in much the same way, this should be of no concern. It does imply, however, that evolution is not the sort of science that can reveal the necessity of what happens, or predict much of what will happen. And recognition of this fact, finally, reveals the futility of much of evolutionary psychology.

3. The Sociological Appeal of Evolutionary Psychology

No doubt the reasons why a scientist is drawn to a particular set of theoretical ideas and a particular programme of research are complex and variable. No doubt also, a central part of those reasons is often, perhaps usually, a conviction that this is the right way to advance understanding. It would, nevertheless, be naive to suppose that there were not other kinds of factors that contribute to the attractions of a particular scientific project, and the plausibility of seeking such factors seems much greater when, as is the case with evolutionary psychology, the epistemological claims seem so defective. One obvious attraction of evolutionary psychology is surely its claims to breadth of application. It may well seem more exciting to make general claims about universal history than to describe the peculiar customs of a tribe in New Guinea or a gang in south central Los Angeles. It is no doubt fascinating to learn, for instance, that '[h]eadhunting is . . . for Ilongots, an act ideally preceding marriage' (Collier and Rosaldo, 1981: 309), but one hopes that one has not thereby learned a general truth about human nature, or anything that will have direct relevance to life in the suburbs. The suggestion that Stone Age females might have been attracted to males who showed their ability to kill other males, and that perhaps this caused a homicide module to be selected in ancestral humans, is much more exciting. Perhaps we also have here the key to inner-city violence among young men in the West and much else. Except, of course, that the evolutionary story is mere speculation. (In fact I just made it up, though I fear I may not have been the first.) One need not be excessively cynical to note that a book claiming to explain why young men exhibit violence has more chance of making the bestseller list than a book describing in detail the culture of the Ilongots.

Whether it is from commendable epistemological enthusiasm, or from the desire to become rich and famous through writing best-selling books, the tendency to exaggerate the scope of the theory with which one is professionally engaged is a familiar aspect of scientific life. I referred to this phenomenon above as scientific imperialism.[7] Physicists envisage final theories of everything, and not surprisingly these turn out to be physical theories. Darwinians, or their camp-followers, make almost equally ambitious claims. Recall the remark from Dennett, quoted above: 'Darwin's dangerous idea is reductionism incarnate, promising to unite and explain almost everything in one magnificent vision.' Economists sometimes claim that all interesting human behaviour can be seen as the attempt by individuals to maximize their utility under conditions of scarcity (an idea to which I return in Chapter 6).

To focus on the application of imperialism to the main topic, it is certainly true that the idea of evolution by natural selection has been among the most fruitful in the history of science. It has made possible, perhaps for the first time, systematic investigation of the origin and nature of the adaptation of organisms to their environments. But even if we treat Dennett's 'almost everything' charitably (in fact over-charitably) to refer only to biology, his claim is massively hyperbolic. To begin with, as critics have made clear for many years, not every distinguishable feature of every organism is an adaptation (Gould and Lewontin, 1979). Of equal importance, the fact that something is an adaptation may point to the explanation of why it is there, but it will often do very little to explain how it works. The complex processes of ontogeny are no doubt broadly speaking adaptations: one cannot reproduce very successfully if one doesn't first manage to develop more or less properly. But this insight does almost nothing to illuminate how development works. To explain this will, presumably, involve serious work in molecular genetics, cytology, embryology, and so on.

Of more immediate relevance to the present topic, the fact that something is an adaptation not only does not tell us how it works, but it need not even tell us what it does. This is for the reason familiar to evolutionists that many parts of organisms evolved for most of their history to do something quite different from what they now do. There is no doubt that the human brain is an adaptation in so far as it is an extremely complex and highly structured organ that plays a vital part in the biological organization of its fortunate possessors. Certain, usually ancient, parts of it have well-understood functions,

[7] See also Dupré (1994).

as those that regulate heartbeat, breathing, and suchlike. But it is a matter of dispute what the more recent and interesting parts evolved to do, and it is obvious that some of the things they do now are not what they evolved to do.[8] Even if the brain really did evolve as the set of modules postulated by evolutionary psychologists, it is not at all obvious that this would give us much insight into how typical human brains now work.

It seems to me that reflection on the scientific backwaters that imperialists so often drift into reinforces a view of science that I have defended in detail on other grounds. Scientific theories are developed to solve problems of fairly specific kinds. Darwin's theory, at least as it subsequently evolved, brilliantly addressed questions about adaptation and biological diversity. The application of the theory then requires very serious attention to the question, 'What is an adaptation and what is not?' This is the reason Gould and Lewontin's critique is of such importance. As in the classical meaning of the word 'critique', it helps to define the scope and limits of a theory. Unfortunately, and perhaps for readily understandable reasons, scientists working within a theory are often more inclined to attempt to expand the scope, finding adaptations everywhere. Exactly parallel is the case of economics. Classical economic theory provided a compelling account of the working of markets. But the conception of a market in which the theory developed was a very restrictive one, and a great deal of subsequent work in economics has attempted to adapt the model to situations that diverge from this ideal case (monopolies, oligopolies, imperfect information, etc.). This attempt to expand the scope of the theory in a gradual way, by relaxing specific assumptions of the central case, is a perfectly legitimate exercise, of course, though opinions will differ as to how successful it has been. This can be seen as a legitimate exploration of the scope of the theory. More recently, however, markets have been postulated that have little more than the loosest analogical relation to the central

---

[8] It is sometimes announced as if it were an important discovery that the brain is a device for processing information. But this is only true because, as it is generally interpreted, it is entirely banal. In the broadest sense of information, in which this is merely a measure of the inverse of uncertainty, and processing information involves no more than a non-random relation between inputs and outputs, then the heart or the liver is an information-processing device as well. More substantive senses of the claim about the brain will be proportionately more controversial with regard both to what it does and to why it evolved.

case: markets in marriage partners, children, leisure, and so on. This is scientific imperialism, and the results, as in the biological case, are generally uninspiring.

Let me finally put this conclusion in a broader context. Wide generality and range of application are often seen by contemporary methodologists as fundamental excellences of science. Some philosophers consider unification of apparently diverse phenomena as fundamental to explanation, and it may seem to follow that the more unification we can find the deeper the explanation we will have provided. But without denying that breadth of application may be an epistemic virtue, it should surely be constrained by the principle that things should not be unified unless they are in relevant respects the same. It seems to me that the search for generality may well prove to be an obstacle to the investigation of at least the more complex and interesting aspects of human behaviour. It may be that such behaviour is sufficiently diverse, and that human cultures and individuals are sufficiently unique, that only a wide range of disciplinary perspectives, applied more through insight than by algorithm, will provide much understanding. And such understanding may necessarily remain at a very local and specific level. I am urging that we should not be seduced by imperialist methodologies simply because we assume that some highly general and wide-ranging theory must provide the correct approach. Current attempts to offer such theories do little to inspire confidence in this assumption.

To return to the official topic of this section, it seems that we are faced with a potential problem in the organization of science. Obvious rewards and attractions accrue to the pursuit of scientific ideas with the greatest claims to generality of application. But if I am right, these attractions do not draw scientists to the projects with the best genuine prospects for contributing to knowledge. Because the institutions for providing rewards to scientists are so much internal to the scientific community, it is very difficult to see what measures might be effective in diverting scientists towards those projects most likely to generate knowledge or, even more importantly, increments in human welfare. And it is not obvious that any such measures should be applied. Though of course I am confident that I am right about evolutionary psychology, I don't believe that I, or anyone else, should have the authoritarian power to shut it down. It would be nice to believe that in the long run some kind of selective process will weed out failed science, but I must confess to being sceptical about this.[9] It is not, after all, obvious to everyone that evolutionary psychology is failing, and the longer it survives the stronger a position its practitioners will be in to define problems in ways such that they can claim to have solved them. The question of how effort is allocated between different possible scientific projects, and withdrawn from those that are misguided or unsuccessful, is a fascinating and important one, but not one I propose to address any further here.

4. Political and Ethical Implications

The controversy surrounding E. O. Wilson's sociobiology embraced from the start accusations not only of scientific inadequacy, but also of objectionable political implications.[10] In a letter to the *New York Review of Books*, signed by sixteen people including most conspicuously Steven Jay Gould and Richard Lewontin, Wilson's sociobiology was characterized as the successor to earlier eugenic theories and Nazism (Allen et al., 1975, reprinted in Caplan, 1978). They wrote: 'Wilson joins the long parade of biological determinists whose work has served to buttress the institutions of society by exonerating them from responsibility for social problems' (Caplan, 1978: 264). Wilson wrote an unsurprisingly angry response to this letter in which he accuses his critics not only of gross distortions, but also of 'the kind of self-righteous vigilantism which not only produces falsehoods but also unjustly hurts individuals and . . . diminishes the spirit of free inquiry and discussion crucial to the health of the intellectual community' (Wilson, 1975b; reprinted in Caplan, 1978: 268). On the issue of political implications Wilson is equivocal. In an article he wrote for the *New York Times Magazine*, he insists that '[g]enetic biases can be trespassed, passions averted or redirected, ethics altered . . . Yet the mind is not infinitely malleable' (Caplan, 1978: 267). The extent to which the mind's recalcitrance will prevent the trespassing of genetic biases, presumably, is to be discerned by further research in human sociobiology.

[9] The importance of such processes has been claimed by several philosophers, most notably Hull (1988) and Kitcher (1993). It is also central to Karl Popper's (1959) well-known falsificationist account of scientific method. The important question is not whether some bad scientific ideas get weeded out (of course they do) but whether science constitutes a reliable instrument for such weeding out.

[10] A comprehensive historical treatment of this controversy is Segerstråle (2000).

In a later attack, this time in the journal *BioScience*, the group of critics has expanded to thirty-five, and named itself the 'Sociobiology Study Group of Science for the People' (Allen et al., 1976; reprinted in Caplan, 1978), to which Wilson again replied in the same journal. The major part of the Science for the People paper is a much more detailed development of the scientific criticisms of sociobiology. The group also sketch their alternative view. Although this point is missed or ignored by Wilson in his reply, they explicitly attempt to distance themselves from environmental determinism with the claim that both genetic and environmental determinisms are incoherently individualistic, whereas 'the individual's social activity is to be understood only by first understanding social institutions' (Caplan, 1978: 288). This is a point I emphasize at several points in the present book. What is more relevant to the present topic is Wilson's elaboration in his reply on the matter of the political implications of sociobiology. Starting from the observation that major political theories must start from some premise about human nature, rights, justice, or the like, he identifies the role of sociobiology as that of validating such premises or, at any rate, as to identify the innate 'censors and motivators in the emotive centers of the brain' (Caplan, 1978: 301). Although insisting that these censors and motivators could in principle be overruled, to do so would be to overrule our 'deepest and most compelling feelings', clearly not something to be done lightly. This should, therefore, 'engender a sense of reserve about proposals for radical change based on utopian intuition. To the extent that the biological interpretation noted here proves correct, men have rights that are innate, rooted in the eradicable drives for survival and self-esteem' (301).

The first thing that I should like to say about these last remarks is that Wilson is to be commended for his forthright acknowledgement that theories about human nature have political implications. This is more explicitly reaffirmed in a paper coauthored with the philosopher Michael Ruse (Ruse and Wilson, 1986), in which the 'naturalistic fallacy', the alleged impossibility of deriving normative conclusions from factual premises, is attacked, and the basis of ethics in evolved biology expounded. It is, at any rate, a commonplace that no normative political philosophy can get off the ground without making some assumptions about what humans are like. Even the assumption that we are not innately like anything, or that we are infinitely malleable, is an assumption. But second, this realization gives scientific claims about human nature a particular importance, and should make us extremely wary of such claims when they lack adequate epistemological credentials. These facts should surely leave Wilson unsurprised that views on such matters claiming full scientific credentials should engender strong feelings. Wilson is hardly the first to note that a strong commitment to innate individual rights would stand in the way of utopian proposals for radical change.[11] Marx, in the *Critique of the Gotha Program*, criticizes what he considers the bourgeois notion of equal rights as standing in the way of his more utopian ideal: 'From each according to his ability, to each according to his needs' (Tucker, 1978: 530). But of course Marx did not think of the notion of rights as referring to an innate fact about humans, but to a central concept of bourgeois ideology. Since some, at least, of the members of the Sociobiology Study Group of Science for the People were openly Marxist (for example the biologists Richard Levins and Richard Lewontin), it is hardly surprising that they should be irritated by the implied claim that their political commitments had been scientifically refuted.

Although this debate certainly went beyond the bounds of polite scientific interchange, it did so in the explicit knowledge on both sides that important issues were at stake. Sociobiology's successor science, evolutionary psychology, has been much less forthright on this point. Contemporary evolutionary psychologists spend very little time discussing the possible moral or political consequences of their conclusions, perhaps because this strikes them as an unseemly activity for a respectable scientist, or at least something only to be done on Sundays, but officially because they have discovered something called the naturalistic fallacy. Unfortunately they have not discovered what this is, or that it is highly debatable whether it is a fallacy, or even, in some cases, what it is called.[12] As far as the matter of what it is, the popular wisdom appears to be that it is the fallacy of supposing that because something exists it must be good (e.g. Pinker, 1997: 50; Buss, 1994: 16; Thornhill and Thornhill, 1992:

---

[11] Though the point is not uncontroversial. Noam Chomsky is a prominent example of a radical progressive who does hold a commitment to innate individual rights. I thank Adrian Haddock for this point.

[12] Thornhill and Thornhill (1992) refer to the 'Naturalist Fallacy', which I take to be the error of supposing that one could learn anything about biology by observing the behaviour of living wild animals.

407)—certainly a fallacy, but not one, as far as I know, that has ever been entertained by a serious thinker.[13]

The term 'naturalistic fallacy' was introduced by G. E. Moore (1903) to refer to the attempt to derive conclusions about what ought to be from premises exclusively about what is in fact the case, though it is more generally associated with Hume's famous remarks about the impossibility of deriving 'ought' statements from 'is' statements (Treatise, III. i. 1). In the case of Moore the thesis was embedded in a now largely discredited ethical theory; and in the case of Hume there is serious doubt whether he in fact intended these remarks to endorse what was later to be called the naturalistic fallacy. But these issues need not detain us since they, as indeed the naturalistic fallacy, are largely beside the point.

No one, I imagine, has ever supposed that whatever is ought to be. The Holocaust happened, but it was not a good thing.[14] But to establish the irrelevance of evolutionary psychology to ethical or political matters would require the almost equally absurd thesis that no factual premises could occur in an argument to a normative conclusion. This would rule out rather uncontroversial arguments such as: murdering people is bad; Joe murdered someone; therefore, Joe did something bad. The point of all this is just that in the sorts of arguments that might plausibly be taken to involve evolutionary psychological claims as premises and ethical or political statements as conclusions, normativity is likely to be found in some other premise. One obvious premise that comes to mind is one to the effect that a good political system will be one that tends to make people happy or satisfied. And what will make people happy or satisfy them is surely the sort of thing that will depend crucially on claims about universal human nature. If people have a mental module that causes them to derive pleasure from collecting and hoarding round shiny stones, then a political system should do its best to provide as many people as possible with access to round shiny stones.

---

[13] It is true that some philosophers, such as Leibniz, have thought that this must be the best of all possible worlds. But this need not be taken as denying the simple sense in which certain things that happen—disease, poverty, violence, and so on—are bad, only as insisting that all in all things would be worse without them.

[14] Here, perhaps, one should exclude Leibniz; though it would have taken all his brilliance to reconcile this disaster with the philosophical principle referred to in the preceding footnote.

Consider, for example, the claim that men have an evolved mental module that causes them, under appropriate circumstances, to rape women (Thornhill and Thornhill, 1992). In a comment on an article defending this thesis I suggested that 'this is not just bad science, it is harmful science'. These accusations are of course connected. As I also noted, 'If such claims really were established, then we would just have to accept these possibly harmful consequences; but because, as far as I can see, the Thornhills say nothing that even affects their probability, the propagation of such claims should be strongly resisted' (Dupré, 1992: 383). In reply the Thornhills fairly brusquely dismissed me as having committed the naturalistic fallacy (at least that is what I think they intended to convict me of; see note 12). The fact that rape is a psychological adaptation doesn't imply 'that rape is inevitable or good' (Thornhill and Thornhill, 1992: 407).

I certainly don't think that if the Thornhills' claim were true it would imply that rape was good, and probably not that it was inevitable. So why is this theory potentially harmful? There are at least two kinds of reasons: first, ill-founded theories of rape will distort our view of the phenomena and so very likely lead to inappropriate policy responses; second, though sociobiological stories do not *legitimate* rape, they will surely affect our moral attitudes to it in ways that are very likely to be inappropriate. To explain the first point requires a little more context. Rape, as I mentioned in the last chapter, is perceived by evolutionary psychologists as one of a range of sexual strategies. The optimal male strategy is to acquire a lot of resources which will enable him to buy sexual access to large numbers of females. But given the approximately one-to-one sex ratio, and the assumption that women are naturally treated as property by men (Wilson and Daly, 1992), this possibility beautifully illustrates the Marxist thesis that property is relational: one man's riches entail another man's, or several other men's, poverty. Those men without the resources to purchase women will fall back on the alternative strategy of raping women. Rapists, therefore, will turn out to be unsuccessful, marginalized men. This patently reinforces a stereotype of impoverished, probably unemployed and criminalized men lurking in alleyways in insalubrious neighbourhoods to prey on passing women. (I leave the reader to decide whether the man thus stereotyped may also have predictable racial characteristics.) And while no doubt there are such men lurking in insalubrious neighbourhoods, this drastically misrepresents the overall sociological reality. A great many rapists are not poor, and a great many rapists are not reproductively opportunistic strangers but friends and family members. Since rape by strangers in alleyways is much more likely to be reported there are probably no reliable statistics on the relevant proportions. But it seems almost certain that strangers in alleyways remain a minority of cases.[15] Thus the biological story contributes to a common misperception of the phenomenon. Of course this would not be an argument against telling the story if it were true, since in that case no misperception would be thereby promoted. Indeed, if it were true it would make it clear that the only way to eradicate rape would be to eradicate poverty. But though eradicating poverty would be an excellent idea, since there is no reason to believe the story true, there is little reason to suppose that this would also eradicate rape, though surely it might reduce its incidence. Unfortunately, since *Homo Darwiniensis* is also competitive and self-interested, it is unlikely that the ones with the resources to acquire all those women will readily concur with such an egalitarian 'utopian' project.

Even more important is the second issue, whether evolutionary story-telling actually serves to legitimate rape in any way. Certainly it does not entail that we should legalize rape, or declare that it is a good thing. If this needed further argument (which it doesn't) we could note that from the point of view of the men pursuing the preferred sexual strategy, rape is a serious property crime, so that even from an evolutionary psychological point of view those in power have every reason to discourage the practice. Indeed, it might even provide an argument for increasing the penalties against rape, since the deterrent force of possible punishment might be seen as having to overcome the force of an innate tendency. Nonetheless, it would surely change our attitude to rape in significant ways. Explaining a behaviour as the direct consequence of an identifiable bit of neurology makes a difference. If my arm flies out wildly and knocks your Ming vase to the ground you will react differently if you think this is

[15] One of the problems here is that sociobiologists typically do not take much account of the difficulties in defining the phenomena they attempt to explain. It seems to be largely assumed that rape is perpetrated by strangers (or ducks) lurking behind the bushes. What is entirely ignored is the extent to which the nature of rape is a matter of continual contestation. Not long ago marital rape was conceptually impossible. And date rape, hardly recognized until quite recently as a problem, now appears to constitute a large proportion of all rapes. The project of fitting this dynamic and contested concept onto some supposedly preexistent biological universal is of course a hopeless one.

a deliberate act of destruction than you would if you knew I had a brain disorder that occasionally caused my limbs to flail uncontrollably. (And differently again, if you thought it was carelessness or clumsiness.) It might be necessary to lock someone up if they were discovered to have a hyperactive rape module, just as it might be best to prohibit the limb flailer from spending much time in china shops. But clearly this is a move away from the treatment of a fully culpable criminal act.

Finally, as well as encouraging certain reactions to rape, it discourages others. It might be that promoting attitudes of greater respect towards women would have a beneficial effect on the incidence of rape, for instance. I don't claim to know how this might best be done, or even whether it would have the suggested effect. But it is the sort of line of investigation that would be suggested by views that see rape as reflecting social attitudes, and that would seem unpromising from the perspective of evolutionary psychological theories. Here I am thinking particularly of the widely held view that rape is not primarily an expression of the sexual instinct at all, but rather an act of violence. My point is not to endorse a view about the cause of rape, but just to emphasize that how one thinks of the aetiology of a phenomenon such as rape has serious consequences for how one responds to it. And so the promulgation of unsupported theories masquerading as science can only be damaging to attempts to deal with social problems.

It would have been easy to make the same arguments about several of the topics most central to contemporary evolutionary psychology, especially to the evolutionary psychology of sex. Also of great potential importance to social matters are sociobiological theories on modules for exchange and accounts of the nature and limits of altruism. These will figure again in a later chapter on economic ideas. My main point for now is just that claims about fundamental aspects of human nature have profound social and political implications, and to deny these implications by hiding behind the tattered fig-leaf of the naturalistic fallacy is naive. And to promulgate such portentous claims when the empirical and theoretical basis for them is, to say the least, tenuous seems to me to be quite properly considered irresponsible.

In the interests of politeness it will be best not to pursue explicitly the extent to which the political consequences of evolutionary psychology provide motivations for its practitioners to pursue it. On the whole, as has been clear at several points in the foregoing, the conclusions tend to be nasty. And as is clear in principle, the conclusions tend to be conservative. The existence of a psychological module cannot be established unless the behaviour it is 'designed' to produce occurs with some frequency, so if evolutionary psychology is to be possible at all the kinds of behaviours that actually occur must frequently be the kinds evolutionary psychologists say Mother Nature has designed for us. Though the fact that we are designed to behave a certain way doesn't make it right to do so, it provides a prima facie reason for saying that it would be best to permit or even promote that kind of behaviour. The emphasis on explaining behaviour in terms of immutable aspects of people will always tend to show that changing behaviour will be difficult, and hence, other things equal, undesirable.

   I don't know whether evolutionary psychologists are typically more conservative than appropriate comparison groups. Certainly good and bad, radical and conservative projects have been proposed and implemented on the basis of evolutionarily grounded beliefs about human nature. If evolutionary psychology were true, even though it would have some tendency to argue for conservative social policies, it would also be good to know its truths if we were attempting to produce social change. And no doubt many evolutionary psychologists believe the naive appeals they sometimes make to the naturalistic fallacy, and think that scientific fact has no bearing on ethical or political matters. Still, theories of human nature do have such bearing, and it is hard to believe that some scientists are not drawn to this kind of research because they find the consequences congenial. But this is a lot less important than is being clear what those consequences are.

5
Kinds of People

1. Introduction

In the preceding chapters I have argued at length against the new biologism that sees evolutionary speculation as a key to revealing the universal constituents of human nature. But I have also tried to avoid embracing the other side of the sterile dichotomy between nature and culture and to keep hold of the obvious truth that the development of a human being involves a long and elaborate interaction between genes and a great deal else both physiological and environmental. In this chapter I shall make some suggestions about what an account of human nature that took seriously such an interactive process might look like.

One way of pursuing such a project is to project the interactive perspective back on to human evolution. One of the more strikingly curious features of evolutionary psychology is the atavistic conclusion that human behaviour is to be understood by reflection on the conditions of life not in the present but in the Stone Age. This conclusion derives from the idea that evolution is ultimately solely the accumulation of favourable genes,[1] and the processes of gene accumulation are extremely slow. Empirical inquiry, on the other hand, suggests that human behaviour has changed dramatically over even the last few tens of thousands of years. Contemporary inhabitants of London or Tokyo would not easily adjust to the mores of life on the African savannah. The sociobiological resolution of this conflict is to conceive of modern humans as profoundly maladapted—apes in skyscrapers. And no doubt this chimes to some degree with a perennial intellectual temptation to see human existence as irremediably

---

[1] One further reminder: this idea is sometimes introduced as a definition of evolution. In my broad usage of the term 'evolution' it is a substantive thesis, and not a very plausible one. Of course, if the narrow definition is insisted on, the question becomes one of justifying the view that evolution has much relevance to contemporary behaviour.

unsatisfactory or even doomed in some such way as this picture suggests.

But this temptation should be resisted, and empirical reality taken more seriously. One of the advantages of the developmental systems perspective (described in Chapter 2, section 4) is that it enables us to escape from the obsession with the genetic that has characterized many recent accounts of evolution. Suppose we think instead in terms of human developmental cycles. As I have already emphasized, human development draws on a great variety of resources, many of which are provided socially or culturally. From this perspective there is no temptation to imagine that human evolution largely ended in the late Stone Age. The developmental cycles by which human generations are reproduced are in fact massively different even from those that obtained a few centuries ago; and those that currently occur in Southern California are very different from those that occur in the forests of New Guinea. From this point of view human evolution is occurring with great rapidity. And this is true both for evolution within lineages, and also in the production of new human lineages that are to some extent evolving independently of one another.

The idea of cultural evolution has been in the air for some time, and precursors have existed at least since the last century.[2] It is plausible that processes analogous to the evolution by natural selection studied by biologists will work to favour differentially some cultural variants over others.[3] Technical work on the processes by which cultural variants might be selected shows clearly that the results of such processes can diverge substantially from those predicted from traditional theories of natural selection. Although this work has clearly advanced our understanding of evolutionary possibility, it also carries with it dangers. Specifically, there is a temptation to treat biological and cultural evolution as almost wholly distinct processes. And this, in turn, promotes the common view that for the human

---

[2] Contemporary discussions derive particularly from Cavalli-Sforza and Feldman (1981) and Boyd and Richerson (1985).

[3] Like any other scientific idea, this one can be overdone (perhaps most are overdone). Richard Dawkins (1976) introduced the term 'meme' to play a role in cultural evolution analogous to that played by genes in biological evolution. Now there is an industry of meme theories ('memetics', even); see Blackmore (2000). One problem with memetics is that genes don't play the role in evolution that Dawkins claimed they did, so the analogy is an unpromising one.

species biological evolution ended in the not-too-distant past and was superseded by cultural evolution.[4] What seems to me to be true is that in the last few tens of thousands of years cultural evolution has become an increasingly powerful force in human evolution, and one that has generated some very important divergent evolution within the human species. The reason it is important to resist the formulation in terms of the replacement of a biological process by a cultural process is that this implies that there is something important that remains constant after this change has occurred, namely biological human nature. This, I think, is a central error of evolutionary psychology. Human nature, on the contrary, has been evolving at an unprecedented rate in recent millennia. And the reason for this is that the forces of natural selection that existed since the emergence of life have been supplemented in human evolution by increasingly powerful forces of cultural evolution.

There is no more to human nature, I want to claim, than the developmental cycles that currently constitute human life. It may be that certain developmental resources including genetic ones have remained constant for some time. But the humans that these genes have helped to build are dramatically different creatures from the ones that more or less the same genes were helping to build a few thousand years ago. Only an obsessive preoccupation with the internal, structural, components of things could make us insist, despite these massive changes, that some basic nature had remained constant. In this chapter I shall explore from various perspectives the processes of cultural change and the nature of cultural diversity. The final two sections will address the value of this diversity.

2. The Power of Culture

It is a familiar observation that culture can produce changes much more quickly than natural selection can bring about the evolution of physical structures. This is because the former depends on learning

---

[4] In an earlier paper (Dupré, 1987b) I came close to endorsing such a view. Someone who has been aiming for a long time to develop more integrated theories of biological and cultural evolution is William Durham (1978; 1991). However, as I argued in the paper just mentioned, it still seems to me that Durham's account puts too much weight on the tendency towards optimal states defined in purely biological terms.

and (leaving aside some worries about learning that evolutionary psychologists make much of) there is no obvious limit to how much learning can happen anew within a single generation. This observation commonly coexists with a countervailing intuition that culture is a weak and ephemeral force compared with the iron grip of the genetic. While there may be something to the view of cultural influences as ephemeral, an inevitable corollary of their ability to evolve rapidly, I want to suggest that far from being (comparatively) weak influences, there is good reason to think that they are often much more potent than genetic influences on behaviour.[5] There is a perfectly straightforward reason why cultural forces are more potent in determining the fine structure of human behaviour than are internal, biological forces, namely that culture is typically normative. Since there is a tendency to think that rules are made to be broken, proven by exceptions, and so on, whereas causation reflects the iron rule of law, this point may be less than obvious, and deserves some elaboration.

Even if, as I do not myself believe, causation occurs only in accordance with exceptionless laws, it is universally acknowledged that laws apply only *ceteris paribus*. That is to say, the statement of an exceptionless law will require reference to a (perhaps open) list of things that could potentially interfere with its exercise. But if we imagine, to take a now familiar example, a brain structure that is capable of causing a man to commit an act of rape, it is perfectly clear that other things will not generally be equal. The point does not depend on any possibly contentious views about the power of culture or freedom of the will. Even in the fully deterministic world of the traditional physicalist, a piece of brain will only have its effect in collaboration with innumerable other internal and external influences. Many of these will surely be other brain pieces or states that will have developed in response to a host of variable conditions.

---

[5] I must confess that I am unhappy with the metaphor of forces employed in this paragraph. For of course it is only a metaphor to depict culture as pushing people towards kinds of behaviour. The metaphor misleads if it suggests that there is some effect that culture (or genes) would have if only genes (or culture) did not interfere. For neither genes nor culture can affect anyone's behaviour without the other. Perhaps a less objectionable terminology would be to speak of the different explanatory resources provided by reference to the genetic or the cultural. But this would have its own drawbacks in suggesting an instrumentalist interpretation that I do not intend. Despite its dangers, I think the metaphor employed should be harmless in developing the points that follow.

Consider, by contrast, the rule—or, in fact, law—against raping people. This is not a causal factor that, in some happy conjunctions of circumstances, will abort an incipient act of rape. It is an unconditional prohibition, supposed to operate wholly regardless of the surrounding circumstances. Unlike killing, there are no circumstances under which raping people is considered justified.[6] Doubts about the mental capacity of the perpetrator may provide some kind of mitigation, though only secure the doubtful benefits of medical rather than penal incarceration. (One of the main concerns raised by sociobiological fantasies about the evolution of rape is that this kind of mitigation might become much more common.)

Of course, rules or laws are seldom or never completely effective. For a variety of reasons they may be violated or, in some cases, overruled. Some rules, such as those enforcing speed limits, are largely ignored. None of this, however, undermines the contrast between a cause, that works only if all the attendant circumstances are correct, and a rule that applies unless some specific factor overrides it. When a rule is not commonly overridden and is considered legitimate and appropriate by a community, it will often be extremely effective in determining behaviour. The parallel that may still seem compelling is not that with an isolated causal factor, but with a piece of machinery. Machines are designed with great efforts made to ensure that other things are equal and interfering forces are excluded. And perhaps the evolutionary psychologist's rape module is to be understood precisely as a piece of machinery?

I will not rehearse here the arguments for doubting that there are machine-like generators of particular kinds of behaviour. The point is rather that rules often do function in an almost machine-like way to determine behaviour from the outside. Think of the countless people driving on the correct side of the road, stopping at traffic lights, eating with cutlery, and so on. Such behaviour is, of course, done unthinkingly, and it is done so largely because any thought would immediately confirm that there was no reason for violating the rule. The conscientious rule-follower, when confronted with a difficulty in following the rule, may go to some trouble to conform nonetheless. It is tempting to see this as closely parallel to the kinds

---

[6] It used to be quite widely held that a woman's 'provocative' dress or behaviour provided, if not justification, at least substantial mitigation. I take it that this view is now not widely defended.

of mechanisms that ensure that a machine works even when some unwanted factor threatens to interfere. In some cases social rules achieve conformity less easily, by imposing sanctions or occasionally rewards. Rules are successful, presumably, to the extent that they provide individuals with compelling motivations for acting in conformity with them; and perhaps most successful to the extent that such conformity is seen as something that should be done simply because there is a rule.

3. Cultural Change: Anagenetic Evolution[7]

A few hundred years ago the ancestors of contemporary Europeans lived very different lives from those enjoyed by their descendants today. Few today could distinguish between the holding of land by free-alms or sokage, sergeantry or knight-service, villein tenure or freehold, matters of enormous importance to the thirteenth-century Englishman. Such things were important, of course, because they determined how people could behave, how they spent their lives. Nowadays there are no villeins, knights, or, in the relevant sense, sergeants, and a good deal of the behaviour characteristic of these roles no longer exists either. Now people drive on motorways or travel by plane, surf the internet, eat TV dinners, and engage in countless other activities unheard of even a hundred years ago. Brains that grow up surfing the internet, it is safe to say, turn out significantly different from those that developed in preparation for a life of villeinage.

The point of these banalities is just to indicate that only someone tightly in the grip of a theory could refrain from drawing the obvious conclusion that humans have been evolving rapidly over the last several centuries. The relevant theory in question is of course the theory that only to the extent that changes are recorded in the genetic endowment of the species should they be counted as true evolutionary changes. But I hope that earlier chapters have undermined the motivations for this gene-centred view. In the absence of such a

---

[7] The term 'anagenetic' is used in evolutionary biology to refer to evolution within a lineage. 'Cladogenetic evolution', the cultural variant of which will be discussed in the next section, refers to the processes involved in the splitting of a lineage into two.

theory it is obvious that changes in behaviour are every bit as important as changes in physiology or molecular structure, and that behaviour is not rigidly determined by genes. The observation that behaviour changes much more rapidly than the genetic endowment of the species then shows that there is more to evolution than changes in the genetic endowment of the species.

Of course, the study of these changes is not normally conceived as part of the study of evolution, but as the domain of human history, especially such branches as cultural and social history. Such a disciplinary distinction is no doubt justified sufficiently by the very different methodology at work. History is studied through the examination of artefacts and most importantly textual artefacts. And very sophisticated techniques have been developed specifically for the analysis and interpretation of such artefacts. It is, indeed, part of my thesis that recent human history is in many ways a unique process differing in fundamental respects from prehuman evolution. But the processes that humans underwent prehistorically have surely not ended. And even if they are usefully distinguished from more recent historical processes analytically, they are surely not so distinguishable ontologically. Cultural and biological evolution are intricately and inseparably intertwined in the history of our species. When, for instance, a lover seduces a sexual partner with rhetoric or poetry, some of the most sophisticated capacities of modern humans are making their contribution to the evolutionary development of the species. Ultimately, human evolution and human history are the same thing.

4. Cultural Species: Cladogenetic Evolution

While explanation of change within a lineage is an important part of the task of a theory of evolution, more fundamental still is the explanation of synchronic diversity. Diversity exists between the members of a species at a time and between different species. Since biologically speaking there is no serious doubt that *Homo sapiens* should be counted as a single species, it might appear that human diversity should be treated entirely as a problem of the first kind. However, it is very natural to divide a consideration of human diversity into two problems corresponding roughly to these two categories. Replacing the concern with differences between two species, however, is the question of differences between cultures. We have, then, three levels at which humans must be understood:

1. biological universals
2. cultural differences
3. individual differences.

With regard to the first, my argument to this point has been that while there are surely causal factors biologically common to all non-defective humans, the interaction of these with innumerable contextual factors makes it unlikely that such factors will in any interesting ways determine the details of human behaviour. And there is no reason to think that there are universal biological features of humans directed at the production of specific modes of behaviour. Rather, such universal features must be seen as more or less constant inputs into the complexly interactive processes by which human minds develop. No doubt genetic factors determine that the large majority of humans will have serious interests in food and sex, for instance. But the behavioural manifestations of these interests will still be highly varied. The fact that a substantial proportion of humans have a predominant sexual interest in other members of their own sex, a fact unintelligible from a crudely adaptationist point of view,[8] should provide a compelling ground for recognizing the profound influence of factors outside the narrowly adaptationist biology assumed by evolutionary psychologists. More likely than the existence of such biological universals is that there are almost universal culturally evolved human universals that might, for reasons discussed above, have a more or less direct influence on behaviour.

This suggestion invites a misunderstanding that must once more be formally disavowed. I don't mean to imply that culture might universally write instructions on a blank slate. Behaviour results from an interaction between culture and biology. Cultural evolution is the evolution of cultural factors that have the capacity to elicit certain kinds of behaviour from creatures with human brains. My earlier discussion of the normative power of culture implied that humans are the sort of beings that are significantly disposed to behave in accordance with rules. I shall try to develop a rather more sophis-

---

[8] I take this unintelligibility to be unaffected by the stories sociobiologists have from time to time made up on the subject.

ticated account of this disposition in the final chapter of this book. For now I note only that such a disposition is not an aspect of any particular culture, but a more abstract precondition of the possibility of there being any culture or any cultural evolution. In a more sophisticated account one might note that the extent to which violations of rules are accepted or punished is an important cultural variable. But for now I shall eschew sophistication. As sociobiologists are fond of pointing out, the capacity for culture evolved. And that, in turn, implies that the capacity for culture was almost certainly biologically advantageous to our distant ancestors. A look around the planet at the present time suggests it may very well be biologically advantageous to their descendants. But none of that implies that culture must take any particular form. And in fact cultures can now evolve to some degree independently of one another, a fact proven by the observation that they can and do evolve in divergent directions.

As we have seen, evolutionary psychologists insist that there are deep underlying commonalities between all human cultures. But superficially, at least, there is a great deal of cultural diversity. Some cultures, mainly now in the surviving tracts of tropical rainforest, have been largely isolated from most of the rest of humanity for millennia, and provide the examples of exotic culture that grace the pages of *National Geographic* magazine and traditional anthropology journals. But cultural difference does not depend on isolation. Catholics and Protestants in Northern Ireland are anything but isolated from one another. On the other hand members of one sect will sometimes report that they grew up without any significant personal contact with members of the other. While it is very much to be hoped that these two cultures will merge, for many years they have coexisted with substantially different customs and presumably a fair degree of reproductive isolation. Similar coexistences occur in many parts of the world, often with tragic consequences. In the United States there is a strong tendency for people of African origin to live in the same areas, speak recognizably distinct dialects, attend the same schools and churches, wear distinctive clothes, and so on (though within this broad generalization many finer and cross-cutting distinctions could also be made). And again, in contrast to some comparable racially identified subcultures in Europe, there is a strong tendency for reproduction to occur within the confines of the culture. The specific constellations of behaviour exhibited by such partially isolated groups can reasonably be said to constitute a specific culture or subculture.[9] Cultural innovations will tend to flow down channels defined by specific cultures, and hence cultures will exist as somewhat integrated wholes.

This leads to the thought that perhaps we should see human culture as constituting a number of distinct cultural species. Of course, cultural species are not fully isolated from one another. I suggested that black Americans form a partially distinct culture in the contemporary United States. But it is a familiar observation that black music is quickly assimilated into white culture. And many cultural innovations, most obviously technological, become available to many cultures simultaneously. So clearly if there are such things as cultural species they are fluid and their boundaries are permeable. The significance of this concept will be easier to evaluate after some reflection on what a traditional biological species is.

5. Humans and Other Species

Biologists have generally assumed that despite frustrating difficulties in defining exactly what constituted a species, the existence of biological species was an objective fact about nature. The existence of a certain number of species was taken to be a matter of fact, as was the membership of any particular organism in a particular species. Identifying these species and assigning organisms to them formed part of the task of the science of biology. This task was inherited from pre-Darwinian biology. An assumption dating at least from Aristotle was that the key to carrying out this enterprise was the discovery of the essences of different species. Possession of the essence, or essential property, of a species was what made an organism the kind of thing it was. So once species had been distinguished in terms of their essential properties, the ability to assign organisms to the right species would follow automatically.

---

[9] I should emphasize that I do not identify the inner-city African-American subculture, even to the extent that it forms a coherent single culture, with a racial group. It is no doubt difficult for a white person to be assimilated into this culture, but certainly not conceptually impossible. And it is surely not the case that middle-class black professionals in racially mixed suburbs belong to this culture. Race, which should of course be defined sociologically rather than biologically, is a factor that contributes to the maintenance of some cultures, but the connection is contingent.

But as many thinkers realized, Darwinism undermined the search for the essence of any particular species: it was difficult or impossible to see how a species essence could be reconciled with an evolutionary process that allowed the transmutation of one species into another. However, this did not in any way threaten essentialism about the species category, about what it was that made a group of organisms constitute a species. And the dominant tradition of answers to the so-called 'species problem' can be seen precisely as offering an answer to the question, 'What is the essence of a species?' Though the pre-Darwinian explanation of the existence of species has of course been rejected, the conception of the problem to be explained was transmitted almost unchanged. More recently, however, it has become more plausible that the problem itself must be questioned. And this is because it is becoming less clear that there is any well-defined unitary phenomenon, the existence of species, for which an explanation is required.

This scepticism about the objective reality of species tends to seem very strange to those with some experience of natural history. When one explores the natural environment in one's immediate vicinity, one will generally find sharply differentiated types of organisms. Problems, however, begin when the exploration is extended in space and in time. Given the fact of evolution, it follows that as one moves back in time one moves towards different forms. If we start with two related forms, and trace their ancestors back in time, we will gradually work our way back to a common ancestor. We will therefore have found a series of intermediate forms joining the related forms. Since it appears that all species are related by descent, the same procedure can be applied to any pair of forms, however distantly related. Thus in the totality of space and time it appears that there is a continuous array of gradually and smoothly differentiated forms, and thus that the drawing of boundaries between forms would appear to be inescapably arbitrary.

So much follows simply from the fact of evolution and the assumption that evolution is gradual rather than saltatory. But perhaps the only conclusion to draw is that we cannot expect the species concept to apply across evolutionary time. In a relatively small time scale, nevertheless, we might still expect sharp and objectively defined species. The problem that arises next is that movement through space will also provide a continuum of forms not found in a local investigation. An extreme example is the well-known case of herring gulls and lesser black-backed gulls, which appear as distinct, reproductively separated, species in Europe, but are connected by a sequence of intermediate forms as one circumnavigates the globe at the latitude in which they are chiefly found.[10] Again, we can respond by limiting the application of well-defined species to a more or less limited spatial as well as temporal extent. The problem is that even then there is no generally applicable criterion for demarcating one species from another.

The classical idea that inevitably succumbed to the Darwinian revolution was the idea that some intrinsic feature of an organism constituted it as a member of the species to which it belonged, the classical doctrine of biological essentialism. Essentialism of this kind cannot even make sense of the idea that one species could evolve into another. There have, to be sure, been recent attempts to define species in terms of intrinsic properties of their members. But these have tended to be explicitly nominalist,[11] a response to the seeming lack of objective criteria of species membership rather than a denial of that lack. Searches for objective criteria of specieshood have been sought elsewhere, in relations to other members of the species. Most familiar of these is the idea that a species consists of a set of organisms reproductively linked to one another, and reproductively isolated from all other organisms. This idea, generally referred to as the 'Biological Species Concept', gained the status of orthodoxy largely due to the influence of one of the leading evolutionists and systematists of the twentieth century, Ernst Mayr. But though it has some important virtues, one of which I shall return to in a moment, it has ultimately proved inadequate.[12]

Problems with the Biological Species Concept (BSC) begin with its limited applicability. It has no obvious relevance to asexual species. Since sex, in its contemporary form, evolved some billions of years into the history of life, it has no relevance to a large part of the

---

[10] The story is actually rather more complicated than this simple if standard summary, though the point it illustrates remains valid (see Mayr, 1970: 291–2).

[11] That is, they have denied that there are objective categories to be discovered. A notable instance is the numerical taxonomy of Sokal and Sneath (1963).

[12] I have discussed these matters in much greater detail elsewhere (Dupré, 1999b; 2001a), and the conclusions I take to follow from the difficulties with 'solutions' to the species problem are there spelled out. Wilson (1999) contains an excellent set of papers presenting current thinking about species.

history of life on earth. Even today most organisms are asexual. Whether most kinds of organisms are asexual is hard to answer. If one were to attempt to apply the BSC literally one would most plausibly identify each asexual lineage distinguished by any genetic alteration as a species. This would imply vast numbers of highly ephemeral species, and lead to the conclusion that the overwhelming majority of species, as well as organisms, were asexual.[13] But even aside from the asexual organisms, boundaries between what are generally taken to be distinct species are not always impermeable to reproduction.

Flowering plants are one of the groups of organisms most prone to hybridization. (It is perhaps no accident that Mayr was an ornithologist. Birds, mammals, and reptiles are the groups in which reproductively determined species boundaries have proven most robust.) One of the classic illustrations of the limitations of the BSC has been the example of American oaks (Van Valen, 1976). Here it seems that despite continual hybridization and the production of fertile hybrids over long periods of time, distinct species have maintained their integrity, apparently showing that reproductive isolation is not necessary for the maintenance of distinct species. One can bite the bullet here and insist that all this shows is that where we thought we had several monotypic species, it turns out that we have only one polytypic species. But this has some unfortunate consequences. First, it begins to divorce the identification of species from the business of developing a useful classification, a departure from what has been the main point of recognizing species for millennia.[14] Second, it will still further reduce the number of sexual species, limiting the applicability of the BSC to a tiny fraction of the species that exist.

The BSC represented the attempt to base a theory of the species directly on what were thought to be requirements posed by the theory of evolution. As the BSC has declined in influence, so attempts have been made to relate the species even more directly to evolution. Phylogenetic accounts of the species treat species simply

---

[13] This point has been made by Hull (1989: 109).

[14] Even where there are no sharp morphological boundaries within large, diverse, and more or less interbreeding groups of plants, the interests of classification may require the specification of somewhat arbitrary borderlines. The conflicts between the aims of classification and the theoretical considerations underlying the 'species problem' are discussed in detail in Dupré (2001a).

as segments of the phylogenetic tree. In its most rigorous version, cladism, the species (and for that matter any higher taxon) is required to be monophyletic. That is, it must include all and only the descendants of some ancestral population.[15] My main objection to this, which I shall not develop here in any detail (but see Dupré, 2001a), is again that it loses touch with any sensible view of the pragmatics of classification. Stylized representations of phylogenetic history (e.g. cladograms) might give the impression that lineages regularly split into two discrete descendant lineages at some fairly regular interval. The reality of course is rather messier. Lineages split all the time. Most branches remain distinct for a relatively short time and either become extinct or re-merge with the remainder of the lineage. When we reconstruct the past history of life we identify only lineages that remained distinct for a fairly long time before either going extinct or (more rarely if they were genuinely isolated for a substantial time) re-merging with other descendants of the ancestral stock. Only these, though of course not all of these, are susceptible of identification. In the end every lineage will meet one fate or the other. And the vast majority will meet extinction sooner rather than later in evolutionary time. Between a monogamous pair all of whose offspring die without issue, and a small population that turn out to be the ancestors of the vertebrates or the flowering plants, there is every possible degree of distinct lineage. The vast majority of these are small and insignificant. The perspective of the past from the present filters off all but the most successful lineages. But of course the taxa we perceive from the present also include within them innumerable unremarked and insignificant lineages that didn't amount to anything much. When we taxonomize the presently existent organisms we lack the kind of filter with which we perceive the past. We do not know which of the lineages that exist now will amount to anything. Our selection can only be based on what seems important to us now for perhaps a variety of more or less theoretical reasons. Given all this, it becomes doubtful whether, from the perspective of classification, there is any reason to worry about such esoteric matters as monophyleticity.

---

[15] Less demanding versions of phylogenetic taxonomy employ a concept of monophyly that requires that a taxon contain only the descendants of an ancestral population, but not that it contain all those descendants.

6. Cultural Species Again

Returning finally to human kinds, our interests here are not classificatory but ontological. My objections to reproductive or phylogenetic accounts of the biological species were based on the belief that we need a general-purpose classification of organisms for reasons of which evolutionary or phylogenetic considerations are only one. We need such a classification simply for the purpose of storing and communicating the vast quantities of information of natural history. No such taxonomy seems needed for humans, and indeed attempts to define such a thing have generally been deplorable.

The relevance of the foregoing discussion of biological species to the human species lies elsewhere. Two points are central. The first is that culture, just as evolutionary change through the biosphere, flows unevenly through human populations, and this uneven flow allows for the existence of vastly more cultural forms than could exist if society were entirely homogeneous. It is hard to imagine what it would be for there to be no biological diversity, as organisms depend for their existence on other kinds of organisms. At any rate, the vital point that underlies the attempt to define species in terms of reproductive barriers is that it is barriers to the flow of evolutionary change from one population to another that make the extent of biological diversity actually found in nature possible. Similarly with culture. The understanding of human culture requires the exploration of the processes that allow cultural items to be transmitted within a group of people but much less readily from one group to another.

The second point is that culture, like biological adaptation, comes in integrated blocks. Cultures are at least partially integrated wholes. Elements of culture must be explained (in part) in terms of the way they mesh with other elements. To take an example almost at random from the ethnographic literature, the widely studied Ilongot head-hunting is part of an articulation of gender conceptions that are 'functioning aspects of a cultural system through which actors manipulate, interpret, legitimise, and reproduce the patterns of co-operation and conflict that order their social world' (Collier and Rosaldo, 1981: 311). This practice is neither a brute inexplicable fact nor something fully explicable in terms of the (somehow) optimal benefits it provides in isolation from this cultural context. Cultures are not, on the other hand, impervious to change or to the influence of other cultures. When we move from the largely isolated and autonomous cultures of contemporary hunter-gatherer tribes, to the subcultures of modern societies more or less complexly interconnected with the wider life of those societies, we are clearly dealing with something highly mutable and composed of parts that some members may reject. Nevertheless some degree of integration will surely be characteristic of anything worth calling a culture, and the persistence of a cultural item will always owe something to its role in that integrated context over and above any benefits it may be seen to provide in isolation. This echoes one of the most fundamental difficulties with crudely adaptationist biology.

But while these analogies are important, the patterns of cultural flow are of course very different from the patterns of biological evolution. Humans in modern societies belong to many different cultural groups and different parts of their behavioural repertoire will be influenced by membership in different such groups. Age cohorts provide cultural barriers that can be as impermeable as those between ethnic or economic groupings. Children attending a culturally diverse school may behave in relatively homogeneous and predictable ways during the school day, but return to very different contexts and different behaviour in the evenings. When people become golfers, members of the Conservative Party, Seventh-Day Adventists, vegetarians, and so on, they will partake, for parts of their lives, in distinct subcultures with distinct characteristic behaviour.

The other fundamental difference between cultural and biological processes is that these cultural barriers are always more or less permeable. Although, as I have noted, the boundaries between biological species are not as absolute as is often assumed, they are often effectively impermeable. And biological evolution does, in the end, create irreversible separation. There may be no telling whether the various genetic streams among contemporary oak trees will eventually diverge, converge, or go extinct. But only human scientists will ever mix the genes of a potato and a snowdrop. Cultural evolution, on the other hand, does not seem ever to produce barriers that are impermeable. Or perhaps this is an illusion generated by our inability to think outside the biological concept of lineage. Perhaps there is no more cultural flow between survivalists in Montana and liberal Jewish intellectual culture in New York than there is genetic flow between pigs and dandelions. The descendants of members of these cultures may certainly move from one to another. But that is just to note that the spread of culture is something quite different from the parent–offspring transmission that is fundamental to biological evolution.

It is not my goal to describe in detail the network of such systems through which culture flows. The most important point is just that biological and cultural evolution *are* quite different processes, and that culture exists and evolves in forms that are quite autonomous from underlying biological processes. This is not, needless to say, to imply that the two are causally isolated. An analogy that should make this clear is with the relation between biological evolution and chemistry. Biochemistry can tell us little or nothing about the course of biological evolution, even though at every stage biological evolution depends wholly on a myriad of biochemical processes and is always constrained by what is biochemically possible. The diversity of life shows us that these constraints leave much freedom, even though in another mood we are greatly impressed by the universality of the genetic code or the discovery of sets of genes that do the same things for bacteria and humans. Evolution, as is empirically obvious, can create an almost unlimited amount of diversity from this relatively homogeneous material.

Just the same things should be said about the move from human biology to human culture. Human biology provides the substrate on which human culture builds, and there could be no human behaviour without the biology that makes humans able to behave. And that biology, finally, is more or less similar for all humans. Here we encounter again the metaphysical vision that informs this work. Central to this vision is the insistence that the entities at many different levels of organization have principles of behaviour and development that are not merely consequences of those at lower structural levels. It is, in other terms, the emphasis on form as of equal significance to matter and the rejection of the tyranny of the microscopic. It is my contention that the opposing metaphysical vision is what gives unjustified credence to projects such as evolutionary psychology and their insistence on understanding human behaviour as best explained in terms of the transactions of microscopic bits of physical stuff.

7. The Value and Future of Cultural Diversity

Cultural diversity is a contentious issue at the moment, and having sketched a view of its nature and origin it seems desirable to say something about its value. It seems to follow naturally from what I have been saying that cultural diversity should be valued and protected just as increasingly many people are insisting we should value biological diversity, and for essentially similar reasons. Just as biological diversity provides an enormous library of chemical and behavioural possibility, so does cultural diversity provide a library of the possibilities for human behaviour and belief. And as surely as the destruction of tropical rainforests and other as yet less than fully exploited habitats is wiping out biological diversity, other forces of modernization are wiping out cultural diversity. In the case of the most isolated of human cultures this is concomitant with the habitat destruction just mentioned. But more omnipresent is the globalization of business that increasingly treats the inhabitants of Berlin and Beijing, Moscow and Mexico City, or Seoul and Seattle to the same ever more homogeneous set of commodities, entertainments, and systems of belief.

This process, however, is much less uniformly deplored than the parallel reduction of species diversity. One reason for this is easy to sympathize with: with cultural diversity comes ethnic cleansing, civil war, racism, and genocide. If we had just one global culture we would be rid of one of the major causes of human strife and misery. But there are also affirmative ideals that induce enthusiasm for the globalization of culture and the elimination of cultural diversity. There are two main categories of such ideals. First, there are straightforwardly ideological considerations. These might involve simply the idea that all cultures should move towards an optimal form. More often, however, they also involve a conviction as to what that optimal form is. The second category consists of economic or commercial arguments for the desirability of cultural homogeneity. I begin with the second.

The commercial arguments I have in mind are of various kinds, some of which will be touched on in the next chapter. One strand of thought derives from the alleged benefits of free trade in maximizing the exploitation of comparative advantage and economies of scale. The increasing power of multinational corporations and the influence of the bodies such as the World Trade Organization that speak on their behalf makes this an influential if not intellectually very compelling set of ideas. Its exponents contend that economic advantages will accrue to everyone to the extent that cultures become, in certain respects at least, more homogeneous. Cultural differences are a bad thing, on this view, to the extent that they interfere with the maximal flow of commodities between societies. The argument tends to be intellectually weak in that it has little to say about what constitutes an economic advantage and how, if at all, these should be weighed against other advantages such as, most obviously, the maintenance of patterns of behaviour and consumption peculiar to a traditional culture.

The argument is given much greater force when supplemented by an account of human well-being in terms of access to the widest possible range of choices. Suppose that people have as part of their (individually variable) natural history a set of ranked preferences, and the higher up the ordered list of preferences they are able to reach the better off they will be. Then, first, the economic argument is supplemented by an account of well-being that seems to entail that greater efficiency of the kind promised by more open trade must improve people's well-being. And, second, any barrier to the free flow of goods, beliefs, or cultural possibility can only reduce the range of possibilities available and thereby leave some, at least, unable to reach what would otherwise have been a more preferred alternative. The most sophisticated development of an argument of this sort is that found in John Stuart Mill's *On Liberty* (1859).

Some of the assumptions underlying this line of argument will be questioned in the following chapter. For now I only point to a familiar paradox. Although at first sight it is plausible that giving everyone access to the products of every culture will increase the choices available to everyone, empirical reality overwhelmingly reveals the opposite.[16] Coca-Cola is not just an additional option for all the people round the globe that drink it, but an option that tends to force other options out of existence. And what counts as competition often ends up as a meaningless struggle between massive corporations selling identical products in different packaging. No one's quality of life is significantly improved by the ability to drink Coke rather than Pepsi. Again a biological analogy is obvious and compelling. One does not increase biodiversity by removing barriers that limit the movement of species. One will get a lot of rats and rabbits everywhere, very probably, and a great many 'less efficient' creatures

---

[16] An especially illuminating analysis of some of the ways this happens is provided by Scitovsky (1976).

will go extinct. Introducing rabbits into Australia did not, in the long run, give native hunters a greater number of species to pursue, though perhaps the Australian vegetation was, for a time, being more efficiently converted into meat. Thus, finally, it is an open question whether removing barriers to trade will provide people with new options that they prefer, or rather remove pre-existing options that they would have preferred to the best that is now available to them.

For better or worse, it seems likely that economic homogeneity will continue to increase, and I don't claim any particular authority as to the extent to which this will reduce cultural diversity or as to the desirability or otherwise of such a reduction. I hazard a guess that if cultural diversity is not to decline it will be because of its ability to survive in subcultures much more integrated, in certain respects, with the surrounding cultural streams than has been true of traditional cultures. Perhaps this will enable cultural diversity to exist in ways that do not so readily generate the horrors with which cultural conflict has been associated in the past, which would surely be an evolution to be welcomed. And there is some encouragement here from the parallel with biological evolution. As I have remarked above, biological diversity seems capable of being maintained without effective isolation between forms. At any rate, the strength of the economic argument in question here will remain undecided.

I turn now to the ideological arguments. Often these have been presented in terms of natural selection: the fittest culture will survive and drive others to extinction. And fitness is good, so this process of reduction of diversity, the elimination of the less fit, is itself a good thing. This argument, even when slightly refined, is unconvincing and perhaps morally repugnant. Fitness, meaning the ability to drive rivals to extinction, is not good. More interesting are arguments that directly defend the superiority of what is increasingly the dominant human culture. Though most people find the self-evident superiority assumed by the Victorian Englishman somewhat ludicrous nowadays, the superiority of Western culture is by no means a view that has disappeared. The most conspicuous contemporary variant emphasizes the superiority of science to the systems of belief in all non-Western cultures, and might reasonably be referred to as 'Imperialist Scientism'. To this topic I turn in conclusion of the present chapter.

8. Imperialist Scientism

A forthright recent statement of scientistic imperialism is Norman Levitt's book *Prometheus Bedeviled* (1999). Levitt became famous, or notorious, for his role in the so-called 'Science Wars' as a scourge of the humanists and social scientists so presumptuous as to speak disrespectfully about science.[17] Levitt thinks science is unequivocally a Good Thing. In contrast to the beliefs of primitive peoples and the teleology and mysticism of believers in alternative medicine, organic farming, or astrology, science is True. The signal and unique contribution of Western civilization is to have hit upon a way of finding out the truth about the natural world, and thus to transcend earlier teleological and false views of the world. This makes Western civilization fundamentally superior to its predecessors and rivals, and makes it desirable that its views and practices should supplant these others.

I don't want to disagree with all of this. I agree that science is on the whole a good thing, and that it is a good thing that can potentially benefit parts of the world other than those in which it arose. The real problem with Levitt's thesis, and the reason why it would be so disastrous to attempt to follow his implied policy of replacing non-scientific with scientific belief systems wherever possible, is that he doesn't make any effort to tell us what science is. Much of the point of the present work is to argue that some of what passes for science is not at all a good thing, and for the kinds of questions that these questionable sciences attempt to address it seems to me entirely possible that approaches not currently thought of as scientific may prove to be far more successful. At any rate, the policy just mentioned can make no sense without a very clear notion of what is and what is not, in the properly honorific sense, science.

The distinguished twentieth-century philosopher of science Imre Lakatos once remarked that scientists typically understand science about as well as fish understand hydrodynamics (1978: 62, n. 2). It is difficult to judge whether this jibe can fairly be applied to Levitt, as he makes almost no effort to explain what he thinks science is. Indeed, it might be that his view of science is best captured by American Supreme Court Justice Potter Stewart's immortal reflection on his failure to define obscenity: 'I know it when I see it.'

---

[17] See especially Gross and Levitt (1994).

He quotes with approval a remark by the geneticist Steven Jones: 'all [scientists] use a shared grammar that allows them to recognize their craft when they see it' (cited in Levitt, 1999: 14). One might immediately wonder whether this doesn't beg the question who is to count as a scientist and therefore to be capable of this infallible recognition, though in theory we could imagine a closed set of people all of whom recognized every other member of the set and no one else as being real scientists. But of course this is not how things work. As with obscenity, there are some hardly disputable examples (sexualized violence against children; molecular genetics), some questionable cases (Anaïs Nin's erotica; evolutionary psychology), and some cases clearly outside the domain (*Sesame Street*; Tarot reading). Levitt has a modest ideal of scientific literacy, that people should know who the experts are and how to obtain their advice (1999: 187–8). But apart from telling us that these experts are the scientists, we are given little guidance on this important matter. If a government plans to reform family law should it consult an evolutionary psychologist or a feminist-leaning psychologist, for instance? Levitt, to be sure, has an unconcealed hostility towards feminism, but absent an account of what makes science scientific we should treat this as mere prejudice.

Levitt does, of course, have quite a bit to tell us about science. It happened in Western Europe in the seventeenth century (1999: 144), and for the first time in human history began to offer truth about the world in place of wishful teleology. And, as I have noted, it is a Very Good Thing: 'Science is the one area of human experience that constitutes, on the whole, a vast, almost unqualified, epistemological success' (287–8). Perhaps so. But not, one hopes, by defining science to include only the successful bits. He does also have some quite unargued philosophical views about science, notably that science assumes monism and reductionism, ideas which I have discussed critically earlier in this book. There are occasional mildly promising suggestions that the excellence of science has something to do with a concern with evidence. This gives rise to the somewhat ironic reflection that Levitt, while constantly presenting himself as one of the misunderstood and beleaguered fraternity of scientists, is in fact a mathematician, the academic discipline that perhaps has less to do with evidence than any other; and certainly, as he concedes (144), not a discipline that started in the seventeenth century, or even in the West. At any rate, and even allowing some special pleading for mathematics, this will not get us very far without some idea of what exactly is done with the evidence. Is science to be distinguished from history, a discipline much more concerned with evidence than, say, population genetics or arguably even theoretical physics? Levitt does even admit that there is no 'readily formulated "scientific method"' (14), but fails to notice that this presents an extremely pressing problem for the proposal that ultimate epistemic authority be ceded to the practitioners of this method.

My own view is that there is no one scientific method, and that scientific practices are extremely diverse. The activities of, say, an experimental high-energy physicist, a beetle taxonomist, an artificial intelligence researcher, and a neuroscientist are prima facie utterly different, and the differences survive closer examination. The project of epistemic evaluation cannot rest with the naive, if not fatuous, claims that Science is Good and I know it when I see it. It requires, rather, careful elaboration of a range of epistemic virtues and consideration of whether particular investigative practices meet up to some or all of these. Such virtues might include such things as empirical adequacy, fruitfulness in generating subsequently confirmed predictions, coherence with other things we take ourselves to know, and perhaps simplicity or elegance. Orthogonal to such evaluations, but no less important, are questions about the benefits (or harms) of pursuing enquiries of various kinds. No doubt the development of such a scheme would show the least disputed exemplars of scientific excellence to display many or all of such virtues, and the practices which Levitt most detests—astrology, quack medicine, and so on—to show few if any. I share enough of Levitt's conception of common sense to allow that generating such core judgements would be a condition of adequacy on such an evaluative programme. But its importance would be to help us to say something about the very many parts of what is widely taken to be science but whose epistemic excellence is less obvious.

Returning, finally, to the question that led me to the discussion of scientific excellence, I hope it is clear that my pluralistic view of epistemic excellence, in contrast to Levitt's monistic scientism, leads to the conclusion that cultural diversity, for reasons parallel to biological diversity, should be treasured.[18] Perhaps the highest

---

[18] And no doubt for other reasons, such as respect for the members of other cultures. But such considerations would be beyond the purview of the present discussion.

concentrations of epistemic excellence can be found in the most successful of the sciences. But little can be found in some projects that masquerade as science and much, surely, can be found in areas of the global intellectual map far removed from the scientific centres of excellence. The argument has in fact been made more eloquently than I could hope to imitate by Paul Feyerabend (1978). Feyerabend, developing arguments explicitly parallel to those in Mill's *On Liberty*, displays the arrogance of many scientists' assumption of the absolute superiority of the scientific method, and the ignorance with which they are happy to dismiss rival traditions. In his earlier, and best-known, work, *Against Method* (1975), Feyerabend had already argued at length that there was no such thing as the scientific method, thus making clear the kind of problem later ignored by writers such as Levitt. The allegedly scientific projects that I have been discussing in this book have few claims on our credulity apart from the general plea that they are part of science. Once this is shown to be an empty claim, such projects can instead speak eloquently for the importance of nurturing as wide a variety of different approaches to human behaviour as possible, a variety that may even rival the variety of different human cultures in which that behaviour occurs.

6
Rational Choice Theory

1. Introduction

Evolutionary psychology offers the theorist of the human mind mental modules that set goals for behaviour and propose strategies for pursuing those goals. Human life is, however, complicated. Even if ultimately everything we do is aimed solely at survival and sex, the roads to these ends are many and sometimes arduous. Survival and value as a reproductive partner depend on such things as health, food, and clothing, and on even more general goods important to obtaining these, such as money and education. In the course of daily living humans must, therefore, decide among an array of possible actions that will variously affect their attainment of these ultimate or intermediate goals. Although evolutionary psychology makes evolution a much more immediate cause of behaviour than arguably it has any right to, it cannot go all the way to an evolutionarily determined action. Some other story is needed to relate evolved goals and ends to their implementation in action. The obvious candidate is Rational Choice Theory.

If anything were to dominate the social sciences as the theory of evolution dominates the biological sciences, it would surely be rational choice theory.[1] To introduce this theory, and its strengths and weaknesses, we must begin with a more philosophical question, 'How could one explain a human action?' The answer accepted by most philosophers concerned with action is something like this: to explain an action we must provide some end or goal to which it was directed and some belief or beliefs about the relation of the action to

---

[1] If the social sciences are construed broadly enough to include individual psychology, then perhaps the dominant component would rather be a family of studies grouped together under the rubric of cognitive science. Sorting out the various elements of this motley collection and relating them to the present discussion would, however, be a task for another book. Here I shall construe the domain of social science more narrowly and strictly, which will, I think, make the statement in the text fairly uncontroversial.

the attainment of the end. So, for example, we might ask, 'Why did Mary buy Microsoft shares?'; and answer, 'Because she wanted to make money, and she believed that Microsoft shares would appreciate.' There can surely be no doubt that this is a common pattern for the explanation of behaviour. To explain is to make something (more) understandable. In the case of such a belief–desire explanation, we make an action more understandable by making it seem rational: if anyone wanted that and believed that, then that is how they would act; or, at least, that is how it would be reasonable, or rational, for them to act.

This account need not be committed to the view that every human action is susceptible of an explanation of this kind. It may be, for instance, that many actions do not require explanation at all. 'Why is he eating his spaghetti with a fork?' It would be odd to ask this question (in Western societies at least) and even odder to answer, 'Because he wants to convey the spaghetti to his mouth and believes that a fork is a good instrument for doing that.' His eating is not an exercise of rationality but an exercise of competence. In the unlikely event that something needed to be explained (perhaps to a traveller from a chopstick-using culture) the explanation would be that this is how it is done here. That's how decently brought up people eat in this society, etc. The explanation doesn't go to the individual action but to the cultural norm. It is worth emphasizing at the outset of this discussion what will be an increasingly important point later, that sometimes questions about behaviour go to social norms to which an individual simply conforms, sometimes they go to the more specific projects of the individual, and it is only in the latter case that questions of rationality in the sense assumed by rational choice theory even arise.[2]

Rational choice theory adds to the belief–desire model of action explanation the idea that human actions are, other things being

---

[2] There is perhaps an even less rational level where we go straight to biology. Why is he eating? Perhaps the answer is, 'Because people do eat at this time in this country'. But perhaps this isn't the normal time for eating, and the answer is just that he's hungry. But it would be odd to add, say, 'And he believes that putting bread and cheese into his mouth will relieve his hunger'. Does a cow believe that putting grass into its mouth will relieve its hunger? Surely not, and surely the explanation of a cow eating is ultimately the same as of a human eating. (Of course, in the latter case there are many other possibilities that don't apply to the cow—he's trying to put on weight; he's trying to impress people with his sophisticated tastes; he's trying to break the world record for pork-pie-eating; and so on.)

equal, optimal for the agents performing them. In moving from rational action to rational choice, we imagine not just one action that will promote the agent's goal, but an agent choosing from a range of possible actions. The theory then assumes that the agent will choose that action that contributes most to the realization of his or her goals. This idea is the basis of most of contemporary economics as well as a variety of quasi-economic expansions into the explanation of phenomena of a not obviously economic kind.

Rational choice theory, like evolutionary psychology, has strongly imperialistic tendencies. Its more enthusiastic exponents, at least, have presented it not as merely a valuable approach to understanding human behaviour, but as the indispensable tool for approaching almost all areas of behaviour. One might think that, as two imperialistic programmes within the same domain, rational choice theory and evolutionary psychology would be natural enemies. But actually the reverse has turned out to be the case. Economists, as noted above the main exponents of rational choice theory, have been notoriously reluctant to say anything about how people acquire the goals that they try maximally to attain when making choices. Evolutionary psychologists, while replete with suggestions as to the criteria that people use in assessing outcomes, are generally cautious enough to avoid suggesting psychological mechanisms that generate behaviour in an automaton-like way. Thus these two programmes are potentially highly complementary: evolutionary psychology says what people want, and rational choice theory says what they will do in their attempts to get as much as possible of what they want.[3]

There is no doubt that the concept of rational choice has provided an extremely fruitful perspective on behaviour. Probably the area in which this is most strikingly true is in the development of game theory over the last few decades. The understanding of strategic behaviour and the problems of optimal coordination of action indicated by the structure of games such as prisoner's dilemma have been important in political theory, economics, and across the range of the human sciences. They have even had significant effects on our understanding of biological evolution (see Maynard Smith, 1983). And the basic understanding of rational action just sketched is fundamental to substantial areas of contemporary philosophy.

[3] This alliance is explicitly proposed to economists by evolutionary psychologists Tooby and Cosmides (1994).

Nevertheless it has also given rise to some expansionist scientific projects as poorly motivated as evolutionary psychology. I shall consider these failures before returning to the more general strengths and weaknesses of the rational choice model later in this chapter.

2. Homo Economicus

The heartland of rational choice theory is economics. Contemporary economics is dominated by what has come to be called 'neoclassical economics'. Although there are, naturally, important differences and debates within the dominant school (paradigm, perhaps), all the major versions can be seen as centred on two ideas: the rational utility-maximizing individual and the market.[4] *Homo economicus* appears in the world with a bundle of goods and a set of tastes or preferences, and goes off to the marketplace to trade his goods with other similar agents to their mutual benefit.

The *locus classicus* for all this is book 1 of Adam Smith's *Wealth of Nations* (1776). Smith's classic work begins with a discussion of the huge gains in efficiency that can be derived from ever more concentrated devotion of the worker to a single task. One man working all day, he claims, might barely make 20 pins. Ten men working together, each specializing on one small part of the process of making a pin, could make 48,000 (Smith, 1776/1994: 5). But none of these collaborators is likely to need 4,800 pins a day, so they must find ways of exchanging them for things they do need. Fortunately we are all born with the disposition to 'truck, barter, and exchange one thing for another' (14), and hence it comes naturally for us to create a market. The pinmakers can take their pins to the market, exchange them for money, and with the money buy as much of whatever they want as they are able then to afford. Thus will the huge gains in efficiency generated by the division of labour enable everyone to have more of

---

[4] There are a number of heterodox schools in economics that would require discussion in a more detailed survey of the state of economics. Best established are Marxist economics and institutionalist economics. More recently have come feminist economics (there is now an excellent journal, *Feminist Economics*) and the critical-realist school centred round Tony Lawson at Cambridge (England). Lawson's book *Economics and Reality* (1997) is in many central respects compatible with the approach of the present work. And there are other dissenters. However, members of these heterodox schools are generally the first to admit the current dominance of the neoclassical school.

what they want. As Smith noted, the larger the market the further this process can be taken. In the small villages of the Scottish Highlands, 'every farmer must be butcher, baker, and brewer for his own family' (19). Many trades can only be carried on in a large town.

Such ideas have continued and developed to an extent that Smith could hardly have imagined. From Smith's ten-person pin factory we have passed through the great factories of the industrial revolution to the multinational corporations of today with wealth and influence that exceed that of medium-sized nations. Organizations dedicated to maintaining the unlimited extent of markets, such as the World Trade Organization, override the policy decisions of otherwise autonomous states. Although Smith's argument was a compelling one, we may wonder if even in the economic sphere it has been pushed too far.[5] As I shall shortly discuss as well, the application of market thinking outside the strictly economic sphere is even more problematic.

The more formal theory of markets, like the contemporary student of economics, begins with the perfectly competitive market. This is a market in which large, in theory even infinite, numbers of buyers and sellers deal with a perfectly homogeneous commodity and in possession of complete information. Under these circumstances it can be convincingly demonstrated, subject to a few further simple assumptions such as that buyers prefer more of the commodity to less, and sellers prefer a higher price to a lower, that there exists an equilibrium price such that the whole stock of the commodity offered for sale is sold, and that this price is in fact reached and the stock of the commodity is sold. Such an argument, less formally presented than is common in current economic texts, can in fact be found in Smith (1776: book 1, ch. 7). In essence, the idea is just that any price at which more or less than what is offered is actually demanded will create competition either among buyers or sellers that will move the price towards the market-clearing level. If the price is too high to clear the market, competition among sellers to sell their surplus will lower it; if the price is too low, competition

---

[5] It is fascinating but not always noticed that Smith himself was very much aware of disadvantages that accrued to the division of labour, especially with regard to the general degradation of the worker required to repeat the same tedious operation hour after hour and day after day. For discussion see Dupré and Gagnier, 1999: 181–2.

among buyers with unmet desires will push it up. Large numbers of buyers and sellers and perfect information about the prices and unmet demands in the market will ensure that participants can only follow market forces rather than in any way determine them.

So far so good. As the economics student soon discovers, from there everything is downhill. Although there are some important markets that may come close to realizing the conditions assumed by the above argument—perhaps local vegetable markets, and on a large scale, perhaps commodity markets and currency markets—there are many that do not. Information is seldom perfect. In general information is costly and often far too expensive to be worth acquiring. I could, before putting a box of detergent in my Supermarket A trolley, make a quick trip to Supermarket B to make sure the same item was not available there for less. But apart from valuing my time, I would surely spend more on petrol getting there than I would be at all likely to save by finding valuable information. Once Supermarket A has persuaded me to start pushing my trolley through its aisles it has me in a thoroughly subordinate market position. And when I choose from among the numerous detergents on offer therein, I am surely not choosing from a wholly homogeneous set of products, though perhaps they are much more homogeneous than I am inclined to think. Whether one does indeed wash whiter (than what?), or make my colours glow, or whatever, are further matters about which I may well lack information. Even if I had bothered to compare full price lists from all the supermarkets I could easily reach, the number of these is small and they may well be involved in some implicit or explicit collusion that distorts prices above the competitive equilibrium.

Many other markets are much more obviously uncompetitive. The aged Mac on which I am now writing will probably be the last I buy, despite my firm belief that it is a much better designed machine than its competition. Software is harder to find, my university refuses to maintain it, and so on. And when I buy my next PC I do not anticipate spending long choosing an operating system, unless the United States government has by then done some quite draconian things to Bill Gates's empire. There are, I suppose, some unexpected natural monopolies in the information technologies that form ever larger parts of our developed economies.

Economists, of course, have interesting and sometimes successful models for dealing with heterogeneoeus markets, imperfect information, monopoly, oligopoly, and so on. I do not intend to make any negative (or, for that matter, positive) evaluation of their efforts in analysing these various market conditions. What I do want to emphasize from these elementary observations about markets is that markets are highly diverse; and the various features that differentiate markets lead to quite diverse behaviour; and in contrast with the orderly equilibration that makes the theory of perfect competition so appealing, these various imperfections may lead to indeterminate and disorderly behaviour. The point of this is, again, not to discourage economists from attempting to extend their theoretical treatment of markets into these more difficult areas, but rather to note that as the subject matter gets further removed from the central model of perfect competition, the relevance of basic Smithian insights becomes increasingly questionable. Thus, finally, to attempt to apply the concept of a market to aspects of human behaviour far removed from even the exchange of commodities for money is to invite confusion. As I shall now go on to illustrate, confusion is just what many of these projects deliver.

My first example of a project motivated by an extreme enthusiasm for the concept of markets is a study of the epidemiology of the AIDS epidemic. The attempt to treat this issue from the perspective of economics has recently been advocated in a book by the prominent judge, and proponent of an economic approach to human behaviour, Richard Posner, and the economist Tomas Philipson (Philipson and Posner, 1993). They suggest that traditional epidemiological approaches to the subject have been grossly misleading because they have generally ignored the behavioural changes brought on in response to the epidemic. I cannot comment on whether this criticism is justified, but to whatever extent it is, it would seem a serious omission. Philipson and Posner propose to fill this lacuna with economics. Since the spread of AIDS occurs as a result of voluntary actions by individuals, and since voluntary actions are assumed to occur in accordance with principles of rational choice, we should look at what makes the potentially hazardous choices involved in the spread of AIDS rational. Hence they consider 'the market for risky sexual "trades"', and proceed on 'the assumption that [this market] ... is ... much like other markets that economists study' (5). My cursory survey of 'markets that economists study' should already make plain that this is an extremely vague and potentially misleading assumption. Scare quotes round the word 'trades' draw attention to the fact that the discussion is not limited to prostitution, but is used in the 'standard economic sense of an activity perceived as mutually beneficial to the persons engaged in it', which, from a standard economistic[6] perspective, is any human interaction whatever. The risk of infection by the HIV virus faced by traders in the risky sex market can be treated as analytically similar to, for example, the risk of default faced by lenders in a credit market. (Though their analysis is restricted to the sexual transfer of the HIV virus, they suggest that the market in used hypodermic syringes could be treated analogously.)

One should surely worry that this starting point introduces, to put it mildly, some analogies. I suppose that a trip to the local singles bar has something in common with a trip to the supermarket: one may go to either place with the hope of bringing back something that one wants, and the most active singles bars are even occasionally referred to as 'meat markets'. But the analogy is very limited. The 'goods' on display at a singles bar do not come with price tags, and those customers who leave at 2.00 a.m. without the desired product generally do not do so because they have surveyed all the price tags (or even shadow price tags) and found that the prices are all too steep.[7] And needless to say, many 'risky sexual trades' arise in situations even less market-like than this. Of course an economist might reply that I have simply ignored the relevant definition of a trade just cited, that of any activity perceived as mutually beneficial by two consenting parties, and inappropriately assumed a naive everyday sense of trade. But unless the preferred definition just means, quite vacuously, something that two people both agree to do without any overt coercion, it must at least involve the picture of more or less careful assessment of costs and benefits (suitably discounted by estimated probabilities). And this seems inappropriate as an account of the 'contract' for risky sex hastily drawn up at a singles bar even more than for more traditional routes to sexual intimacy. The more fundamental question is whether such an abstract account of human interaction is of much use outside its core application, in this case the buying and selling of commercial goods. It is this sceptical question

---

[6] For my use of this term, see Chapter 3, n. 7.

[7] It is often said, for example, that as closing time approaches customers remaining in the bar start to look increasingly attractive. This increasing value of the product is not, however, guaranteed to produce a market- (or bar-) clearing sequence of transactions.

that casts general doubt on the desirability of this imperialistic approach to human behaviour.

Philipson and Posner arrive at some surprising conclusions from their economic perspective. Most striking is the following. Policy responses to the HIV epidemic have generally advocated and subsidized widely available voluntary testing. The implicit assumption supporting such a policy is that someone aware that they are HIV infected is less likely to endanger others by engaging in behaviour liable to spread the disease. Philipson and Posner argue that this may well be counterproductive and increase the spread of the infection. A trader contemplating risky sex will estimate the benefits of the activity and subtract the costs of contracting a very unpleasant and ultimately fatal disease, discounted by the probability of contracting the disease. The latter will depend on the prevalence of infected persons in some relevant class and the likelihood of contracting the disease even from an infected person, both of which are quite low in most circumstances. But a person who has been able to determine, by means of a reliable test, that he or she is already infected will have no chance at all of contracting the disease (just as the only people whom it is absolutely impossible to kill are the already dead). Thus for such people the HIV-related costs of a sexual transaction are zero, and they will be much more likely to engage in risky sex, thereby spreading the infection. Philipson and Posner do acknowledge that there may be some altruists in the population, people for whom infecting a sexual partner with a fatal disease would involve some disutility. If we think of an altruist, in this context, as someone for whom the knowledge that they were HIV positive would decrease the likelihood of their engaging in risky sex, then since the proportion of altruists is unknown, the actual effects of widely available voluntary testing remains an empirical issue. Nonetheless, it is clear throughout the book that Philipson and Posner are strongly inclined to believe that the effect will be deleterious. (Even altruism can have perverse effects. Since altruists, who would prefer not to kill their sexual partners, will generally be less active in the risky sex market, the market will have a higher proportion of egoists and therefore be all the more dangerous.) Unsurprisingly, this fits into a pattern of scepticism about government interventions of various kinds, a pattern characteristic of devotees of the generally beneficial consequences of unfettered markets.

I do not know of any comparably detailed studies from the point of view of evolutionary biology except for those that focus on the evolution of the virus itself. It is plausible, however, that speculative evolutionary biology could be deployed here in support of the economistic perspective. To begin with, evolutionary biology will certainly treat sex as a major behavioural imperative, while notoriously viewing altruism as a problematic and limited hypothetical motive. Thus biology could readily be deployed in support of Philipson and Posner's evaluation of the relative significance of these motivational factors. Much more speculatively, it has been suggested that a function of the often surprising mate-selection criteria encountered in nature is the selection of mates with relatively low parasite loads (Hamilton and Zuk, 1982). But if there is evolutionary pressure to find ways of selecting relatively parasite-free partners, there will be equally strong selective pressure to conceal the parasites that one has. Thus if, as evolutionary imperialists frequently appear to do, we see human deliberation as a Darwinian enabling mechanism, the behaviour of Philipson's and Posner's egoistic HIV spreaders represents the implementation of a deep biological imperative.

Although economism and evolutionary speculation often support one another naturally—evolution explaining particular behavioural preferences, and economics exploring the consequences of choices in accordance with those preferences—relations are not always so smooth. My second example is of a phenomenon that is surely a great embarrassment to human sociobiologists, the spectacular decline in fertility in recent decades in the world's wealthiest countries. Although heroic attempts might be made to argue that somehow or other, contrary to all appearances, people are continuing to maximize reproductive success, this would have all the plausibility of the thesis that we live in the best of all possible worlds. Evolutionary approaches to this question must at least back away from direct explanation of behaviour towards more causally distant psychological mechanisms. Of course, the claim that people have some serious interest in sex, an interest with an evolutionary basis in the connection between sex and reproduction is hardly controversial. But it is evident that modern Western humans have managed to decouple the interest in sex from its consequences in reproduction, something facilitated by widely available and effective methods of contraception. Evidently decisions about reproduction are now typically made in some much broader context of individual or even social goals, and evolutionary biology appears to have ceased to have any significant implications for human fertility.

Economics may therefore seem much better placed to address this phenomenon. The *locus classicus* for such an attempt is Gary Becker's *Treatise on the Family* (1981; revised ed., 1991). Becker's narrative begins with an account of the marriage market within which men and women attempt to acquire partners with whom they can create utility-maximizing families. Such partnerships having been formed, the members must then determine how to allocate their resources to achieve the highest accessible level of utility. Among the major items that they wish to produce and—in a chillingly Swiftian phrase—to consume, are children. Already we see, in contradistinction to a naive evolutionary perspective, that raising children is just one among many possible tastes, so that we should have no a priori expectations about natural fertility rates. Becker then makes a further distinction between two possible sources of family utility, quantity of children and quality of children. These interact in obvious ways: high-quality children are costly, involving expenditures on such things as health and education, and thus a demand for superior children reduces the demand for numbers of children. Various explanations are offered for the changes in preference between quality and quantity of children with economic development. Whereas subsistence agriculture creates a demand for lots of children as a source of cheap labour, developed economies provide a range of economic opportunities that increase the returns on investment in the production of high-grade offspring. And, remarkably enough, there is a correlation between the level of education of mothers and of their children, though Becker cautions (1991: 153) that this may not be a direct causal connection, but rather a consequence of the lower demand for quantity of children among more educated women, and the inverse relation between the demand for quantity and demand for quality.

I do not want to deny that the economic, and even the evolutionary, speculations just reviewed may have some grain of truth in them, and might provide a legitimate basis for empirical investigations of various kinds. And certainly economists and biologists engaged in the kinds of speculation I have just described will generally be ready to admit that various other factors are also relevant to the phenomena under consideration. So it may be wondered what my objection is. I shall return a little later to some more detailed critical reflections on aspects of Becker's work, but first I shall make some more general critical reflections on this genre of scientific work.

3. Problems with Imperialist Science

I just remarked that there is nothing obviously inappropriate in using economic styles of thought as a source of hypotheses about various areas of human behaviour. And indeed both Becker and Philipson and Posner do occasionally relate the hypotheses they extract from economic arguments to empirical data. However, typical imperialists do not merely establish embassies in foreign countries and offer advice to indigenous populations. And similarly, economic imperialists do not merely export a few tentative hypotheses into the fields they invade, but introduce an entire methodology and one that is in many cases almost entirely inappropriate. Here I mean by 'methodology' two things: first, a set of core assumptions about how to conceive of the phenomenon under investigation, in this case human behaviour; and second, a methodology in the strict sense of a style of scientific argument. I begin with the first of these.

Economics conceives of human behaviour as the exercise of 'rational' choice. A person faced with a decision is conceived as estimating which action will generate the largest expected excess of benefits over costs to the agent. So, for example, in the case of the person contemplating risky sex, the benefits are the expected pleasure to be derived from the sexual contact, and the costs are the estimated risks of acquiring a fatal illness. Against the objection that human motivation is a bit more complex than this, it may be replied that this economistic perspective is at worst benignly vacuous. If it is true that people are motivated by some concern for the well-being of others, or some wish to behave in ways customary among their social group, or according to their conception of morality, duty, etc., then behaviour satisfying these various criteria will simply be seen as providing some contribution to the utility of the agent. 'Utility' here need not be taken as referring to any real measurable quantity, but only as a fictional device for describing a consistent set of preferences.[8] Both the economic works I have been discussing exemplify

---

[8] The normative evisceration of utility, through marginalism, ordinalism, and finally revealed preference theory, is a dominant theme of the last hundred years of economic thought. See Gagnier (2000).

such a strategy. Thus Becker suggests that parents have a utility function in which utility from their own consumption is traded off against the utility of their children (something from which, of course, the parents are expected to acquire utility themselves). Philipson and Posner conceive of altruists as people who derive disutility from communicating a fatal disease to another person. And so on.

I want to suggest that this may in the end be vacuous, but it is hardly benign. Here I touch on a wide range of philosophical work that has criticized the rational choice foundations of economic thinking (see e.g. Sen, 1979; Anderson, 1993). The most obvious point is that to treat altruism, morality, or accepted social norms simply as tastes that some people happen to have—I like candy and fast cars, you like morality and oysters—is grossly to misplace the importance of norms of behaviour in people's lives. Morality is what for many people makes sense of their lives, not just one among a range of possible consumables. Perhaps there are people for whom what primarily makes sense of their lives is the acquisition of cars or oysters. But most of us, I suppose, would consider this pathological, and would not consider that such lives made much sense.[9] Second, whereas in principle the rational choice perspective may be capable of encompassing any mix of self-interested and non-self-interested concerns, in practice it almost invariably reflects the assumption that the former are far more significant. The most prominent contemporary defender of the project of analysing ethics in terms of economically conceived rational choice, David Gauthier, writes: 'It is neither unrealistic nor pessimistic to suppose that beyond the ties of blood and friendship . . . human beings exhibit little positive fellow feeling' (1986: 101). Perhaps this is only a contingent defect of the universalizing rational choice perspective, but it seems a pervasive defect, and as I shall next explain, it is one that may even be causally connected to that perspective.

The economistic perspective on human behaviour is scientistic in a sense that is of particular relevance to, and particularly pernicious

[9] Of course I mean no disrespect for the life of a car dealer or a shellfish cook, say. But the sense in which it is appropriate for such people's lives to focus on cars or oysters is a much more specific one, the sense in which ethics may be the central concern of a professional philosopher. I am thinking of the way in which, at a more abstract level, an ethical vision may be of equal importance to the chef, the ethicist, and even the car dealer. The point is not that everyone has such an ethical vision (though perhaps many or even most people do), but that having one is not at all like having a taste for oysters.

for, the study of human behaviour. It is scientistic in that it conceives of itself as an objective and (ironically) disinterested reflection on the facts of human behaviour. But human behaviour is not an immutable set of phenomena awaiting the correct scientific analysis, but rather is subject to constant historical evolution. And theorizing about human behaviour is always to some extent an intervention in this evolution. Striking empirical confirmation of this claim in the present context is provided by recent research by the economist Robert Frank and his associates (Frank, Gilovich, and Regan, 1993). Frank administered tests of the disposition to cooperate to students before and after completing introductory courses in microeconomics and in astronomy, and found that the economics students were strikingly less cooperative at the end of the course.[10] (The astronomy students were marginally more cooperative.) In confirmation of this result he also discovered that economics professors made significantly fewer contributions to charity than professors in other fields. I share the conviction of Adam Smith (1776/1974: 15), confirmed by Frank's first result, that on the whole people do not select their occupations because of their different dispositions, but acquire different capacities and dispositions as a result of their occupations. An unusually highly refined awareness of self-interest may be harmless or even appropriate for analysing the futures market in pork-bellies, but it does not seem to me an asset for investigating familial relations or the spread of AIDS. I shall consider the issue of the normative connotations of economistic thinking at the conclusion of this chapter.

The point about the effects of theorizing is not limited to this apparently immediate causal consequence. And the effects of economic theorizing are, surely, much wider than merely the pernicious influence of economics courses. Purported knowledge of how societies work will affect how people think about society and social relations and affect the kinds of institutions that it seems appropriate to put in place. Such effects can take many forms. Some are plainly coercive: Third World countries seeking economic aid are generally required to implement policies thought desirable by Western economists, for example. Only a society deeply imbued with eco-

---

[10] The tests involved engaging in prisoner's dilemma-type games with one another. The structure of these games is such that each player can always do better by refusing to cooperate with the other, but the outcome is better for both when both cooperate than when neither does.

nomic thinking could suppose that a market in pollution licences was the best way to control pollution. It may, indeed, very well be the best policy for controlling pollution. But if it is, it is surely so only because we live in a society in which a certain style of economic behaviour (maximizing profits without much concern for externalities) is considered natural and acceptable. Policies of these kinds, finally, undoubtedly have effects on the people who live with them. They will, that is to say, be much more likely to think it natural and appropriate to behave in ways they conceive as maximizing individual utility. And this, finally, will have a tendency to make economic thinking about such a society true. The categories of economics are paradigms of what Ian Hacking (1999) has described as 'looping kinds', kinds whose application has often profound effects on the social phenomena they are intended to illuminate.

My final objection to the substance of economism brings us closer to the more purely methodological aspects of scientism to which I turn in a moment. Although rational choice theory is not microreductive in the same sense as evolutionary psychology, it is certainly part of a broad reductionist programme as exemplified in its typical commitment to explanations of social-level economic phenomena by appeal to individual behaviour. It is, at any rate, reductive in the broader sense with which this book is primarily concerned, namely in its commitment to monocausality: it attempts to explain ever larger domains of behaviour in terms solely of the maximization of the satisfaction of desire in the light of belief. One of the most important omissions from this monocausal scheme has been best captured by Amartya Sen's insistence on the distinction between desire and commitment. In his classic paper 'Rational Fools' (1979), Sen notes that the picture of an individual addressing every decision with a utility-maximizing calculation is impoverished as well as unrealistic. What it most strikingly ignores is the extent to which human action is guided by principles that are not typically subject to being overruled by minute considerations of personal utility. Such principles include, for example, moral and political ideals, conceptions of good manners or good taste, and loyalty to friends, organizations, or nations.

The deeper significance of this objection will be explained in the final chapter, where I shall argue that it is through an understanding of commitment to principles that we can begin to develop a proper conception of human freedom. And it is this conception that most fundamentally opposes the reductive and homogenizing conceptions of human behaviour against which I have been arguing. For now, I merely summarize part of the argument that will be developed there. Reductionist philosophical positions will be strongly inclined to treat principles of the kind just mentioned as fictions or illusions, requiring analysis in terms of more fundamental and robustly physical entities. However, in the context of the non-reductionist metaphysical pluralism that I advocate, there is no reason to adopt such a deflationary attitude.

Humans possess a remarkable range of causal capacities, many of which, as I have argued earlier, depend on social contexts. Which capacities they will exercise in what situations are in part determined by the set of principles they have adopted and internalized. From this perspective, the problem of understanding human behaviour will centrally involve such questions as how general principles governing behaviour come to be socially entrenched, and how and to what extent these become accepted, interpreted, internalized, and acted on by individuals. Once again, what must ultimately be understood is a complex relationship of mutual dependence between individuals and society. Different societies leave very different amounts of space outside social norms for individual choice and, perhaps, personal utility maximization. Familiarly enough, the proper extent of such normatively unconstrained behavioural space is a fundamental dimension of political debate. And one obvious such space is the realm of behaviour which is the proper domain of economics. But, as Kant saw, the most interesting dimension of human freedom is not the ability to maximize individual utility through unconstrained choice, but the possibility of choosing principles by which to determine one's own actions, and principles that may be evaluated in terms of a wider social good. The scientistic projects I have been criticizing in this chapter may be epistemologically attractive in imposing a familiar kind of order on human action. But the price of this epistemological elegance is the obliteration of the most interesting and important aspects of human behaviour.

4. Central Themes in Scientistic Methodology

So much for the impoverished content of economic thinking. The temptation of such impoverishment, I believe, derives in large part from commitment to methodological imperatives, and it is to these that I now turn. 'Serious' economic approaches to behavioural questions do not merely suggest ways in which economic factors might impinge on matters of human concern: it is a professional responsibility to dress up these suggestions in some quasi-mathematical guise. Becker's *Treatise on the Family* (1981/1991) will serve well to illustrate the problem. (Since Becker is a Nobel laureate it is of course no surprise that he adheres to the professional demands of his discipline.) The book addresses matters of obvious importance—the selection of marriage partners, decisions about child-bearing and rearing, and so on—and on occasion has illuminating insights and even empirical data that bear on these topics. On the other hand, the work as a whole is rendered largely unreadable by a continuous and obfuscating veneer of mathematics. I say 'obfuscating' because the mathematical modelling at almost every point requires a level of abstraction that removes the discussion from any serious connection with the phenomena. Thus, for example, the first chapter on marriage markets, which addresses the economics of monogamy versus polygamy, assumes for purposes of model construction that all men and all women are identical. The object of the exercise is to determine how many wives will maximize a husband's income, which is total family income minus the income of his wives. Becker is at pains to emphasize (e.g. 1991: 96) that men are not assumed to value wives for their own sakes, but only for their contributions to family productivity. While this does help the analogy between marriage and the market in, say, cars (or better, oil, since for analytic simplicity the number of spouses is allowed to vary continuously), it also makes it most unlikely that any conclusions will have much relevance to human mating patterns. As the chapter progresses, it is true, the men are allowed to differ in quality (superior men have characteristics that positively affect the marginal productivity of their identical wives), and in the following chapter, on assortative mating, high-quality and low-quality women are also introduced. Again, though, differences in quality are only variations in some set of properties that contributes to efficient family production.

This, I am inclined to claim, is all fairly self-evident nonsense. These abstractions would seem harmlessly ridiculous in a prose discussion of the nature of the family. The problem, however, is that the nonsensicality is heavily obfuscated. As I mentioned, the book is largely unreadable. It is, however, skimmable—informal discussion leads me to believe that even economists usually skim this kind of work—and conclusions can be seen to emerge. There is some danger that such conclusions will be given some weight as appearing from the serious scientific work (see all that mathematics) of a world-renowned scientist. One might even be tempted just to read the quite clear and intelligible summaries and conclusions at the end of each chapter. If one were to do so, one might read, for example, that 'one of the more surprising conclusions of our analysis is that progressive taxes and expenditures may well widen the inequality in the long-run equilibrium disposition of disposable income' (1991: 231). Although many will be happy just to welcome this conclusion, those less enthusiastic about it might go back and see where it came from. In the section on 'Government Redistribution of Income' (1991: 218) we may read, before diving into the alphabet soup, Becker's model for a progressive taxation system. This amounts to a flat rate tax on income, combined with a lump sum redistribution for everyone. Although this does provide an asymptotic approach towards the flat tax rate with increasing income, and thus might strictly be counted as progressive, it is hardly the kind of thing serious proponents of progressive taxation have in mind. This strikes me as rather weakly supporting the surprising conclusion in the summary. Although this is a particularly egregious example, it serves to illustrate clearly enough the rhetorical dangers of the formalized abstractions that fill the pages of Becker's book.

The point of this example deserves elaboration. I am not trying to suggest that there is anything necessarily wrong with idealization and mathematical analysis of idealized models. Indeed, from Newtonian mechanics to the best work in behavioural ecology and the social sciences there are superb examples of the benefits that can be derived from such an approach. Idealization, indeed, may be a prerequisite for any scientific analysis, perhaps even for theoretical thinking in general. The central problem is in maintaining a link between idealized models and claims that are made about real phenomena. The move from point masses and frictionless planes in Newtonian mechanics is empirically validated by the behaviour of real objects sliding or rolling down slopes, and a variety of well-validated methods exist for incorporating greater complexities into more realistic models. Empirical and theoretical bridges between the idealizations of the models and the greater messiness of the real world are an essential part of what makes Newtonian mechanics a paradigm of successful science.

It is the stark opposition to this ideal situation that makes Becker's juxtaposition of elaborate mathematical argument with substantive claims about real phenomena so profoundly unscientific. And it is the Procrustean attempt to conform the messiness and complexity of human life to the constraints of mathematical representation that leads to all the nonsense about identical wives, infinitely divisible children, and so on. Returning to the last example, the claim that progressive taxation 'may well widen the inequality in the long-run equilibrium disposition of disposable income' is, as Becker acknowledges, surprising. It is surprising because its contrary, that progressive taxation may well narrow inequality of disposable income, is self-evident. I don't say that this is a self-evident truth that we should hold onto come what may. But its self-evident status is not even threatened by the fact that an ingenious economist can come up with a model under which the opposite appears to happen. It is not threatened, because to deserve any credence such a model would have to show that for a significant range of plausible policies, for example the progressive income tax systems and welfare expenditures of a range of contemporary governments, and for a range of plausible definitions of inequality, assumptions about economic behaviour, and so on, the paradoxical result obtained. Becker attempts nothing of the sort, and this casual appeal to mathematical modelling, in fact based on an empirically untested (as far as I know) and only marginally progressive policy, shows nothing at all. The essential quality that Becker's model lacks, and which is perhaps the feature that most sharply distinguishes the good idealized models from the bad, is robustness.[11] The more assumptions that have to be made and the more factors that are omitted, the harder it will be to demonstrate that the model is robust—continues to generate the same result—when the assumptions are varied or additional factors are incorporated. It is of course trivial to provide a counter-model that achieves the less surprising result. Imagine, for instance, a taxation system involving a progressive income tax that reached 100 per cent at, say, $100,000, and confiscated all wealth in excess of $1,000,000. The proceeds are to be spent on providing free health,

---

[11] I am grateful to Philip Kitcher (in correspondence) for emphasizing this point to me. Kitcher's (1985) criticisms of early sociobiological models provide exemplary illustrations of the strategy of showing the lack of robustness, and hence merit, of many such models. The idea that robustness is the *sine qua non* of genuinely explanatory generalization is developed in an important recent paper by James Woodward (2000).

education, transport, and other social goods available to all. It would, I think, insult the reader's intelligence to present a mathematical argument to show that if implemented in the United States or the United Kingdom, say, this would reduce inequality in disposable income. Whether it would lead to a massive decline in national income or revolution or whatever are interesting issues but beside the point. My point is only that, as was obvious at the outset, the possible or probable outcomes of taxation policies will depend greatly on what those policies are, and it is most unlikely that there is any uniform connection between a policy being (however slightly) progressive and having a specific effect on equality of disposable income.

I have been suggesting that much of the simplistic content of economistic theories derives from, and even claims legitimation from, a certain view of scientific methodology. A possible response is that the methodology I have been considering just is the scientific method. It may be said to be of the essence of scientific understanding that it requires concentrating on a very small number of factors and treating other factors as fixed. It is also sometimes suggested that a central feature that distinguishes science from prescientific modes of understanding is the commitment to quantitative techniques. And quantification, finally, requires abstraction. To decide to measure one feature of a class of objects is to privilege that feature over others. This combination of abstraction and quantification is characteristic of the modelling techniques found in much of biology and the social sciences, is signalled by the *ceteris paribus* condition on scientific laws, and is perhaps most strikingly exemplified in many attempts at microreductive explanation. Perhaps it is this combination of abstraction and mathematical representation that is most distinctive of the most uncontroversially scientific practices of enquiry. But if this is so, it indicates a limit to the possibility of scientific understanding of phenomena as complex and multicausal as human behaviour. This is the most general moral to be drawn from the deficiencies and absurdities of imperialistic economism. Taking scientific accounts of this simplified character too seriously as explanations of human behaviour exposes us to grave dangers, normative as well as epistemic.[12]

[12] The limitations of mathematical modelling, specifically in physics and in economics, have been discussed extensively and with great insight by Nancy Cartwright (see especially her 1999). My thinking on these matters is indebted to her work and conversations with her over the years.

5. A Comparison of Imperialism in Evolutionary Biology and Neoclassical Economics

As keys to the understanding of human behaviour I have found something a little more positive to say about neoclassical economics than about evolutionary psychology. But that should hardly be surprising. Economics is, from the outset, directed at understanding human behaviour. How successful it is at understanding the aspects of behaviour towards which it is primarily directed is a matter of debate,[13] but it would be unduly harsh to suggest that it has no contribution to make to any area of behaviour. Evolutionary theory, on the other hand, is primarily directed at much broader questions than the specifics of human behaviour; and since it is acknowledged by all that human behaviour involves some considerations absent from the broad extent of biology to which evolutionary ideas apply, it is neither surprising, nor a reproof to the theory of evolution by natural selection, if it has little to contribute to this topic.

There is, nevertheless, one major epistemological advantage that evolutionary theory possesses over many manifestations of rational choice theory: the central concept, fitness, has rather more content than does the quantity maximized in many economistic models. The bugbear of rational choice theory is vacuity: whatever people do must reflect what they most want. Evolutionary biology, by contrast, is hampered by a central concept with rather more content. Although evolutionary biologists have notoriously been accused of making up any old story that will link some observed behaviour to reproductive success, the need for such a story is a non-trivial requirement, and stories can be investigated for plausibility. One cannot simply say that these particular animals prefer service to their kind or a glorious death to reproductive success. Whereas utility for the agent can be attributed to whatever an action delivers to the agent (or at worst to whatever we believe the agent expects the action to deliver), reproductive optimality must often be sought some causal distance behind

---

[13] There have, of course, been a large number of serious attacks on the aspirations of neoclassical economics. Such have come from prominent figures in all of the heterodox schools of economics mentioned in n. 4 of this chapter. I have myself discussed mainstream economics in a more generally critical vein in Dupré (1993b; 2001b); see also Dupré and Gagnier (1999).

overt behavioural performance. This provides room for hypotheses that can be non-trivially confirmed or refuted.[14]

I don't mean to imply that work in the tradition of rational choice theory is universally vacuous. What the danger of triviality rather suggests is that we may need to be very careful in distinguishing the merely formal from the genuinely empirical aspects of rational choice theory. Much of the most impressive such work, for example in the theory of games, surely belongs in the first category. This is particularly obviously so for such applications as evolutionary game theory. As a technique for developing interesting hypotheses this has undoubtedly made significant contributions to evolutionary theory. But whether such models have real application to understanding the evolutionary history of real groups of organisms is quite another matter, requiring very difficult empirical investigation. The tendency to make this transition too readily is a central shortcoming of evolutionary psychology, and it is no surprise that the same fault is often exhibited within rational choice theory. It is characteristic of both fields that their most imperialistic moments are also the moments in which they are most inclined to substitute purely formal demonstrations of possibility for the hard work of demonstrating actuality.

6. Simplistic Economics vs. Sophisticated Pluralism: The Case of Work

Throughout this book one of my underlying concerns has been with the tendency encouraged by a certain conception of the scientific method to take too literally highly abstract models. I propose now to illustrate this tendency with an example that is central to parts of economic theory, the phenomena and theories of work. Work appears in various guises in the theories of economics, but increasingly in the history of economics it is seen as an input to production acquired in the marketplace. There is thus a labour market much like the markets for raw materials in which producers engage in the

[14] This is not to say that the concept of fitness cannot be put to some almost wholly vacuous uses. Mathematical treatments of fitness values in population genetics often achieve a level of vacuity comparable to anything accomplished by abstract rational choice theory. (Or so I have argued, in my 1993a: 131–42.)

attempt to make the most efficient and therefore most profitable use of available resources. What I want to emphasize is that this is only one of a number of crucial respects in which work has been conceived and is still conceived, and that an adequate picture of the place of work in human life requires attention to all of these. The point of this, in turn, is to show the limits of any particular abstract model in illuminating the human condition, and the necessity for a multi-perspectival, pluralistic approach. This is not, of course, to deny any significance to economic models of labour markets, but rather to point to the inevitable poverty of imperialistic approaches that suppose that the economic perspective is by itself adequate to the understanding of human behaviour.

One ancient tradition sees work as, in the words of Adam Smith, toil and trouble. As God said to an earlier Adam: 'In the sweat of thy face shall thou eat bread, till thou return to the ground' (Genesis 3: 19). But this toil and trouble was also, for Smith, the source of all value:

What everything really costs to the man who wants to acquire it, is the toil and trouble of acquiring it. What every thing is really worth to the man who has acquired it, and who wants to dispose of it or exchange it for something else, is the toil and trouble which it can save to himself, and which it can impose upon other people ... Wealth, as Mr. Hobbes says, is power . . a certain command over all the labour, or over all the produce of labour which is then in the market. (Smith, 1994: 33)

The opposite of work, for Smith, is ease, and he later remarks that '[i]t is in the interest of every man to live as much at his ease as he can' (1994: 821). Here, then, we have two central meanings of work: first, a source of discomfort, a thing to be avoided; and second, the source of value, that which transforms what nature provides into what humans desire or need.

The second of these, the so-called 'labour theory of value', is more famously associated with Marx. Although it has been very widely rejected as providing a general basis for a theory of value, there is surely something right about it. The idea that exchange value, or price, could be cashed out in terms of the amount of labour accreted in a product is probably ultimately unsustainable. But the idea that much work is directed at the transformation of nature for the better serving of human ends is surely right; and, more importantly, the idea alluded to in the reference to Hobbes, that wealth is a command over the labour of others, is surely correct as well. This idea, again, was more famously developed by Marx:

> Money's properties are my properties and essential powers—the properties and powers of its possessor . . . I am ugly, but I can buy myself the most *beautiful* of women. Therefore I am not *ugly*, for the effect of *ugliness*—its deterrent power—is nullified by money . . . [The possessor of money] can buy talented people for himself, and is he who has power over the talented not more talented than the talented? Do not I, who thanks to money am capable of all that the human heart longs for, possess all human capacities? (Tucker, 1978: 81)

Labour provides the capacities to meet human needs or wants, but it need not, of course, be the labour of the person whose wants or desires are to be met. Here we move towards what has become the dominant economic conception, labour as a commodity.

But before turning to that there is another important conception of work to be found in Marx, and one that quite contradicts the Smithian idea of toil and trouble and work as something to be minimized. This is the idea, common today only among the professional middle classes, that work provides the sphere for personal self-fulfilment. In *Capital* (1867), Marx was concerned to describe labour 'in a form that stamps it as exclusively human'. In contrast with the sometimes exquisite productions of non-human animals, the human worker

> not only effects a change of form in the material on which he works, but he also realises a purpose of his own that gives the law to his modus operandi, and to which he must subordinate his will. (Tucker, 1978: 344–5)

It is the absence of this connection between the agent's purpose and the work performed that constitutes wage labour as alienated and thus as failing to serve the development or fulfilment of the agent.

> [The worker's] labour is therefore not voluntary, but coerced: it is *forced labour*. It is therefore not the satisfaction of a need; it is merely a *means* to satisfy needs external to it. Its alien character emerges clearly from the fact that as soon as no physical or other compulsion exists, labour is shunned like the plague . . . Lastly, the external character of labour for the worker appears in the fact that it does not belong to him, that in it he belongs, not to himself, but another . . . [I]t is the loss of self. (Tucker, 1978: 74)

In sharp distinction from Smith, for Marx it is only the alienation of labour that makes it into something to be avoided. Another nineteenth-century thinker who emphasized the role of work as a vehicle for self-fulfilment was John Stuart Mill, though rather than toil and trouble, Mill described the thirty-five years of office work he performed for the East India Company as 'leisure' or 'relaxation' from his real work as a writer and public intellectual (Mill, 1873).

So far I have distinguished work as toil and trouble, work as self-fulfilment, and work as the universal source of value in goods, that which is commanded by wealth. The first two clearly represent incompatible categories of work, and perhaps coincide to some degree with Marx's concepts of alienated and unalienated work. The last conception of work focuses on the economic agent as a producer, as someone whose labour transforms some aspect of the world. Since the 1870s, however, economics had shifted its attention from the producer to the consumer. The guiding image is not of the world transformed by human labour, so much as of the individual maximizing personal utility. From this perspective, the capacity for labour is simply part of the endowment with which individuals enter the marketplace—for the less fortunate their sole endowment—which they will then attempt to exchange for as much as possible of the things they want.

The first thing to be said about the labour market, perhaps, is that there is surely no such thing. The market for professors of philosophy has no significant connection to the markets for architects or advertising executives, or, except for the most unfortunate, to the residual market for those with no special skills at all. There are, then, numerous largely distinct though partially overlapping markets (if builders are much cheaper than surveyors I might risk employing a builder to investigate the structural condition of a house I am contemplating buying, for instance). When economists contemplate these markets they often imagine a purchaser buying labour for investment in an enterprise, and attempting to buy labour until the marginal product equalled the marginal cost of the last unit acquired. This may be a useful model for the proprietor of a mine deciding how many workers to send down the pit, but it has doubtful relevance to the negotiations between the AUT and the CVCP.[15]

---

[15] The Association of University Teachers and the Committee of Vice-Chancellors and Principals, the bodies nominally responsible for determining British academic salary scales. I say 'nominal' because of course the parameters within which this negotiation takes place are largely determined by how much money the government allocates to the university sector.

Or a person who needs a lawyer will generally pay whatever lawyers charge if she can afford to do so, and otherwise will do without the benefits of the legal system. How much lawyers charge, in turn, has more to do with their ability to prevent competition through restrictive trade organizations than with the marginal product their services provide.[16] All this is just to say that like markets for goods, markets for labour vary greatly in the respect to which they correspond to the central economic model of the competitive market.

The consequences of these economic processes of wage-rate determination have some relevance to the meaning of work for individuals. Job satisfaction is presumably enhanced by good pay, and decisions to embark on a particular career will often be influenced to a greater or lesser extent by the financial rewards anticipated.[17] But clearly this is very far from the whole story. This much was well known to Adam Smith, who devotes a superb chapter of *The Wealth of Nations* (book 1, ch. 10) to a consideration of the numerous and diverse reasons for the great disparities in pay between occupations. Smith assumes that these inequalities can be explained on the assumption that they reflect the rational choices of individuals. In pursuit of this project, he identifies five causes of such inequalities. First, occupations differ in how pleasant they are. Because '[t]he trade of a butcher is a brutal and odious business' it is generally 'more profitable than the greater part of common trades' (1994: 116). The second consideration is the ease or difficulty and expense of learning a trade. This he takes to account for the higher compensation of those trained in the 'ingenious arts and liberal professions' (118). The third factor is the constancy or inconstancy of employment; the fourth the degree of trust that must be reposed in the worker (we could not place the trust accorded to our physician or

---

[16] See Garfinkel (1981: ch. 3) for a persuasive elaboration of this point.

[17] This was presumably very much the case in the early 1980s when enormous proportions of bright students wanted careers in investment banking that were seen to offer the possibility of enormous wealth. This had the additional consequence of fuelling an enormous boom, in the United States, in undergraduate degrees in economics. It was widely rumoured that the reason for this was not that formal economics had any great relevance to the practice of investment banking, but that anyone who had shown a sufficient determination to make money to put up with four years studying economics clearly had the dedication to this end required for a successful career in investment banking. Paradoxically, the very irrelevance of economics to the profession thus made it more suitable as a qualification.

attorney in someone of 'a very mean or low condition' [121]); and the fifth is the probability of success, the low level of which accounts for the high fees charged by lawyers.

The second, third, and fifth of these factors might be said merely to point to the complexity of calculating the gain to be expected from employment, and Smith incidentally notes that this will probably be done badly.[18] The fourth factor presents a problem that I shall not try to resolve, namely that unless trust is disadvantageous to the person trusted, the market should not permit this premium (this point is noted by Cannan in a footnote to Smith's text [1994: 121, n. 15]). The most significant point, however, is the first. For if Marx and Mill are right, and work is a vital source of self-fulfilment, perhaps even the most important factor in a fulfilled life, then this feature of work should have a decisive effect on the labour market. Smith addresses mainly kinds of work that he thinks will be almost universally agreed to be pleasant or unpleasant. But in a more individualistic age, and with the notion of self-fulfilment on the table, we should surely expect that people will have different tastes in work. Some may agree with Smith that an innkeeper, 'exposed to the brutality of every drunkard, exercises neither a very agreeable nor a very creditable business' (116), while others will see proprietorship of a charming country pub as their highest aspiration. To the extent that such additional costs and benefits of labour are simply parts of common packages, it will be easy enough in principle at least to adjust a standard market model to accommodate them. However, the vision of the individual as both a seller of labour trying to get the best price for what he brings to market, and at the same time a consumer of career paths trying to acquire the career that will best satisfy his particular tastes in work, is much harder to make coherent. This is in part because the first part of this duality assumes that work is a source of disutility—something one is prepared to suffer as a means to things one wants—whereas the second sees it as a source of utility. Clearly it cannot be both, or at any rate it cannot be treated analytically as both at the same time.

This leads immediately to a general source of the weakness of such models. The models are supposed to present an ideal, rational

---

[18] People will irrationally enter the lottery of a profession with small chance of success because of 'an ancient evil remarked by philosophers and moralists of all ages', 'the over-wheening conceit which the majority of men have of their own abilities' (124).

outcome. Goods are distributed to the people who want them most. Why do people want particular careers? No doubt a part of the answer can be found in such things as the odium of slaughtering animals or the griminess of life as a collier remarked by Smith. Smith also suggests that honourable and enjoyable employments are badly paid in monetary terms. This seems logical from his perspective, but seems at least today to fly in the face of experience. Common sense suggests that there are some careers that provide handsome remuneration, honour or at any rate status, and a fulfilling life, and others (not, admittedly, often referred to as careers) that provide none of these. Doctors, politicians, or scholars may often complain that they are poorly paid, but not because they are financially worse off than slaughterhouse workers, gravediggers, or refuse collectors. Though we can all agree that these latter occupations are necessary and in no way dishonourable, it would be difficult to deny that more status attaches to jobs in the former group. It is commonly assumed that people are mainly sorted into these diversely rewarding occupations on the grounds of natural aptitudes. Coming to the marketplace with an innate aptitude for legal disputation, say, puts one in a better bargaining position than if one has only a strong back to offer. It is noteworthy that Smith would have none of this:

The difference between the most dissimilar characters, between a philosopher and a common street porter, for example, seems to arise not so much from nature, as from habit, custom, and education. When they came into the world, and for the first six or eight years of their existence, they were, perhaps, very much alike . . . About that age, or soon after, they come to be employed in very different occupations. The difference of talents comes then to be taken notice of, and widens by degrees, until the vanity of the philosopher is willing to acknowledge scarcely any resemblance. (1994: 17)

And, with a move that should at least enlist evolutionary psychologists to his point of view, he notes that prior to the development of the modern division of labour, everyone must have had the abilities necessary for all or most of the essential provisioning of life.

Smith is surely right about this. Although there is some social mobility in modern societies, there is no dispute that the best way to get a high-status, high-reward job is to have parents with such jobs. And the fact that the ability to conduct abstruse legal arguments about taxation is so much more highly paid than the ability to build walls or to unblock drains surely has more to do with the relations to social power of those who argue in courts than it does to the much more agreeable nature of the activity of unblocking drains.

Let me now summarize the main point of the foregoing discussion. The pivotal concepts in describing human life, not least those that are central to the economic aspects of life, are imbued with many levels of meaning. Economic models can capture some of those meanings and provide useful insight into aspects of the phenomena in question. But the economic perspective,[19] like any other, highlights certain features and obscures others. This is true not just for the narrowly economic perspective, but also for its expansion into the broader vision of rational choice. The choosing activity of individuals is an inadequate basis for understanding behaviour even in the case of an aspect of human life that we are inclined to think of as primarily economic, such as employment. Among the major issues that arise concerning work in substantial independence from processes of individual choice are the following:

1. The ways in which work is seen as providing meaning or purpose to human lives.
2. The ways in which socio-cultural processes award differential status to different kinds of work, and thereby constitute certain occupations as more credible sources of meaning to some lives than are other occupations.
3. The processes that determine the differential material rewards to different occupations. No doubt there is an economic dimension to this, but such phenomena as the massive and well-known differences between the 'compensation' (a wonderful term for students of the theory-laden) of executives in the United States and in Japan make it clear that there are also social particularities involved.
4. The ways in which different people become eligible for higher- and lower-status jobs largely, though by no means exclusively, through their inherited status.

No doubt there are many more such questions. They are, at any rate, historical, sociological, or anthropological questions, and only a

[19] It was a regressive step, in my view, when the study of political economy was reconceived, about the 1870s, as economics *tout court*. The former term much better captures the integration of economic with other basic social phenomena. For details of this transition, see Gagnier (2000: ch. 1).

combination of these various perspectives can begin to do justice to the phenomena. Gould and Lewontin (1979) famously criticized adaptationism by claiming that the constraints under which adaptation occurred were as important as or more important than the selective processes that brought about adaptation. The point here is closely parallel. Market phenomena of the kind that economists and imperialistic rational choice enthusiasts insist on occur within frameworks that are the contingent products of particular histories. These frameworks are as essential in determining what happens as are the economic phenomena. Pure rational choice is as impotent to illuminate human life as is pure evolutionary speculation, and for closely similar reasons.

7. The Normative and the Positive

In this chapter I have moved fairly freely between discussion of a general perspective on human agency, rational choice theory, and a body of scientific theory, neoclassical economics. This requires no apology, as neoclassical economics is by far the most influential expression of the rational choice perspective, and as neoclassical economics becomes more imperialistic it becomes increasingly indistinguishable from a mere commitment to the universal applicability of rational choice theory. In this concluding section of the present chapter, I want to consider what kind of normative assumptions are involved in such a commitment. I shall start by considering this question in the narrower context of economics and then extend the discussion to rational choice theory more generally.

Economics is generally presented as dividing into two distinct branches, positive and normative. There is no doubt that the former is by far the most prestigious. In common with traditional positivism and contemporary scientism, the underlying assumption of this distinction is that there is a set of economic facts and laws that economists are employed to discover, and that what to do with these is largely a matter for politicians or voters to decide. In light of this assumption normative, or welfare, economics has generally limited itself to the question whether there are economic actions that are indisputably beneficial. This results in attention to the criterion of Pareto optimality: an economic allocation is said to be Pareto optimal if there is no possible transfer of goods that would improve the

lot of some agent or agents while harming no one. No doubt failures
to achieve Pareto optimality should be addressed where possible. But
the 'optimality' in 'Pareto optimality' is dubious to say the least. If, for
example, I possess everything in the world and I derive pleasure from
the knowledge that I own everything in the world, this distribution
of goods constitutes a Pareto optimum. If some crust of my bread
were diverted to a starving child, I would no longer have the satisfaction of owning everything in the world, and similarly with any other
possible transfer. But this would be an unconvincing argument that
this distribution was optimal, or even good. There are of course
countless Pareto optima, which by itself suggests something anomalous in the use of the term 'optimum'.

The problem is perfectly obvious. While we can all agree that
Pareto optimality is a good thing if we can get it, the issue of interest
is which of the many Pareto optima we should prefer. Pareto optimality is really about efficiency, whereas we are interested in properly
normative economics in matters such as justice.[20] But it is at least a
perfectly intelligible claim that economics should aim simply to
describe the mechanisms of economic activity and leave it to others
to decide what to do with them. That this is indeed the case is the implicit assumption of positive economics.

I don't want to deny here that this is conceivable (though I do
have serious doubts about this), but I do want to insist that it is an
extremely undesirable conception of the business of economists.
One way to see this is to note that when we consult supposedly
expert economists about what might be good economic policy we
might naively suppose that they would have useful advice to offer us.
But on the conception under review it turns out that, apart perhaps
from pointing to the occasional departure from Pareto optimality,
they have no relevant expertise whatever. They are, after all, experts
in efficiency not policy. But since economists often seem willing to
offer advice on policy, it seems disingenuous that they should deny
that normative questions are part of their discipline. More seriously,
it is quite clear that there is an implicit normative agenda to the vast
majority of economic thinking. Because economists believe they
have something to say about economic efficiency, they are naturally

[20] As noted earlier, the issue here can be traced in part to the change from political
economy to mere economics in the late nineteenth century. The earlier term much more
satisfactorily insisted on the indissoluble connections between issues of these kinds. For
further discussion see Dupré and Gagnier (1999); Gagnier (2000).

inclined to think of this as a good thing. And as the clearest measure of efficiency is the ability to produce more stuff with the same resources, economists are often inclined to think the goal of economic activity is to produce as much stuff as possible. Even if this account of the aetiology of this goal is disputable, it is hard to dispute that many economists do assume this goal; and assuming a goal is a good way of avoiding the vital intellectual labour of considering what the goals of economic activity really should be.

It is, in fact, an enormously difficult question, even if we agree that something should be maximized by economic activity, what that something should be. Not infrequently positive economics assumes that the real question is about maximizing wealth measured in monetary terms, and tragically many politicians seem willing to accept this facile view. An obviously preferable goal would be something like standard of living, except that that would be little more than a marker for the difficult question of what constitutes standard of living. Sen (1987) makes it clear that any satisfactory analysis of this concept will be only marginally related either to any standard account of utility, or to the accumulation of wealth. It is also clear that even if we knew what constituted a standard of living we would still have to face the task of deciding how this should be distributed. Surely the utility of increases in standard of living declines as one reaches more comfortable levels, so greater good can be gained by distributing standards of living more equally. And there is also the question of who should be among the beneficiaries of a distribution. Should we care about the standards of living of foreigners for instance? Do the as yet unborn have any claim on a decent standard of living? Or must we consider the well-being of non-human animals, or the effects of economic activity on the environment?

Still none of these questions shows that it is impossible to take the traditional view of economics as proceeding purely in terms of the pursuit of economic efficiency, and to graft normative discussions on to accounts of the mechanics of economies. But two points make this implausible. First, it is overwhelmingly unlikely that economists will address the questions relevant to producing normatively desirable outcomes unless they have some conception of what those outcomes are. As a matter of fact the questions they address are very often questions that concern increasing monetary measures of outcomes. And this leads to the second point. Experience suggests that economists do devote a lot of energy to considering how to maximize monetary outcomes, and offer a good deal of advice to governments and others in power based on the outcomes of such investigations. Arguably, a great deal of this advice lies behind policies that are extremely harmful to the well-being of people (and other sentient beings).

Adam Smith, once again, explained admirably the advantages of free trade: larger markets allow further division of labour and therefore greater efficiency gains in production. In recent economic policy discussion this seems to have been elevated in many quarters to a fetish. Institutions such as the World Trade Organization, doing the bidding of the United States and other Western governments and with the enthusiastic support of most mainstream economists, attempt ruthlessly to root out any local policies that threaten to restrict the flow of trade. But the negative consequences of this are so obvious to so many that the meeting of the WTO in Seattle in 2000 was the occasion for prominent public demonstrations of dissent. To take an example almost at random, British pig farmers are currently in the process of going extinct, in large part because British law puts significant limits on the maltreatment of pigs. Free trade treaties, however, make it impossible to prevent the importation of pigmeat grown in million-strong cities of porcine misery, with which British-grown pigmeat is wholly unable to compete. Ironically enough, it could be said that this precisely illustrates one of the central intended advantages of free trade, that goods should be produced wherever there are greatest comparative advantages for their production. A large part of the comparative advantage of continental European and American producers over their British competitors is the absence of legislation limiting maltreatment of pigs. Whether the torture of pigs is a reasonable price to pay for cheaper bacon is a question that simply doesn't arise.[21]

A problem more directly related to the measure of well-being is the measurement of development in poor countries. This issue has been superbly described by Marilyn Waring (1988).[22] Waring

[21] Another issue is the potential environmental consequences of a million large animals excreting in a highly congested area. The lakes of pig faeces threatening to engulf the neighbours of these massive and monstrous farms, and the foul stench surrounding them, raise the important question of economic externalities, but a question beyond the scope of this discussion.

[22] I have been told that things have improved considerably since Waring's book, whether or not in partial reaction to that work I do not know. The point remains that definition and measurement of economic variables is a complex matter on which often turn matters quite literally of life and death.

documents the normative force of the assumption that societies were organized into male wage-earners and females at home doing domestic work on the way that economic development was measured with national income statistics. Since female work in this model was a constant, outside the processes of economic development, it did not need to be measured, and the difficult problems of measuring work in the home could be ignored. Economic development would, therefore, be judged to be taking place if the volume of (male) employment outside the home was increasing. Unfortunately in some cases this might involve the employment of men in the production of cash crops on land that had previously been cultivated by women to feed their families. Not infrequently the cash wages would be spent in bars on the way home from work. From the point of view of national income, this was something instead of nothing. From the point of view of human welfare, this was a drastic decline. It has now been realized that the way to feed children, at least, is to make sure as much of the money as possible goes to women rather than men. This anthropological (or for some, no doubt, sociobiological) fact is one of those without which economic analysis is impotent.

The issue is once again the meaning of work. For many economists until recently work has been conceived solely as that which is traded in the labour market. Hence domestic labour is immediately excluded by definition. No doubt another factor operative here is ready measurability. If work were just what was paid for, and the value of work was just what was paid for it, it would be easy to feed numbers into mathematical models and come up with results that appealed to a certain sense of the scientific. But of course much, perhaps most, productive activity is not paid for, and what is paid for it has little bearing on its value to human well-being. Efforts are now made to provide much more inclusive measurements of production, and such efforts are surely to be welcomed. Nevertheless, there can be no doubt that such measurements involve a considerable degree of arbitrariness, and the recognition of this should engender a degree of scepticism about the fetishism of the quantitative common to both much of scientism and much of science.

Even economic measurements that are widely disseminated and that provide the basis for important public policy decisions are far more problematic than is often realized or admitted. An obvious example, well known to economists, is the measurement of changes in the price level. Many people probably suppose that when they are told that inflation has reached 5 per cent some objective fact about the economy has been reported. But in reality this is a fact crafted in ultimately quite arbitrary ways. How are the changes in the prices of computers, potatoes, yachts, rent, and so on to be weighted in providing such a number? A reportedly stable price level resulting from an increase in the cost of food balanced by a decline in the costs of air travel and holiday villas will not be experienced as price stability by those too poor to afford these latter goods. There is, in fact, no non-arbitrary answer to the problem of weighting the various changes in prices of diverse goods. Different answers will have important and divergent effects. Numbers officially accepted as measures of changes in the cost of living will determine changes in index-linked payments, affect negotiation over pay changes, and perhaps motivate changes in interest rates leading to a further cascade of consequences.

Such consequences cannot be avoided despite the realization that there is no objectively correct way of producing such figures. And this, finally, leads inescapably to the conclusion that such a measure cannot be constructed without implicit or explicit commitment to value judgements. Whose cost of living do we want to measure and who do we want to compensate for changes in their cost of living? Policy-makers and the public would surely be better served if economic indices were explicitly advertised as designed to serve particular goals—preserving the standard of living of the poorest segment of the population, for example—than by the illusion of pure objectivity.[23]

---

[23] A fascinating development in this area is the increasing circulation, in the United Kingdom at any rate, of what is referred to as the 'underlying' rate of inflation, a measure that excludes changes in mortgage payments consequent on changes in interest rates by the central bank. Presumably the rationale for this measure is that it is the official task of the bank to control inflation by adjusting the interest rate. Raising interest rates is supposed, after a time lag, to decrease the rate of inflation. So no doubt it seems prudent to exclude from consideration the immediate and substantial increase in the price level consequent on increasing mortgage payments. Perhaps this case is of more interest to students of the theory-ladenness of science than its value-ladenness. Nevertheless, the practice surely serves to obfuscate, for obviously sound political reasons, the fact that a large part of the mechanism by which the bank's action works to reduce inflation is by extracting money from mortgagees, thereby reducing their expenditure and reducing overall demand. To employ such a strategy, rather than, say, increasing income taxes is a political decision reflecting obvious value judgements. And the use of this peculiar measure is surely in part a device for concealing this judgement under a cloak of quasi-objectivity.

The idea of value-free economics is, then, a myth, and very probably a pernicious myth. But the central route I have described whereby values come into economics is through attempts to aggregate and compare the utilities of individuals. Is there nevertheless an individual level at which objective economic facts obtain? Provided my specific tastes remain constant there is, for example, a (more or less) objective measure of changes in the price level for me. Prices of particular goods should be weighted according to the strength of my disposition to spend money on those goods. So is there at least a value-free level of rational choice theory, a level at which an objective account of the behaviour of individuals can be formulated? The safe answer here would be to say that it doesn't matter much either way. My present concern with theories of human behaviour is with their ability to inform social policies, and this requires aggregation of individual psychologies. Moreover, our interest in understanding individual psychology as an end in itself is brought into play most strongly when individual psychology becomes abnormal, and hence when we have least reason to believe that individuals are choosing, in any sense, rationally.

But I am inclined to question more fundamentally the legitimacy of the individual perspective in question here. Certainly individual preferences vary greatly, but values do not exist independently of a social context. Just as a mass of individual decisions creates important kinds of social phenomena (for example, market phenomena), so social phenomena constitute the range of values from which people choose. Perhaps this does not preclude there being a sense in which the choices that are rational for an individual or a group of individuals involved in strategic interactions with one another may be an objective matter. It is interesting to note that the social prevalence of values, for example of cooperation, may change the likely outcome of strategic interactions for the better, even though they are 'less rational'.[24] At any rate, one thing that is wrong with rational choice

---

[24] This is the moral of the Prisoner's Dilemma Problem, and perhaps most visibly in contemporary versions of the so-called Tragedy of the Commons. The latter is essentially a multi-player prisoner's dilemma. The classic example, from which the name derives, is of common grazing land. Everyone is best off if the optimal number of animals grazes on the commons, but each can do better by adding a further animal. Everyone pursues their individual self-interest in this way, and the amenity collapses, to the detriment of all. Just such a process appears currently to doom attempts at international agreement on fisheries policies.

theory as a general perspective on human behaviour is its individualism, the erroneous assumption that social phenomena are no more than the aggregates of individual behaviour. Important ways in which this picture goes wrong are the unavoidable role of value judgements in processes of economic aggregation, and the fact that even what seem superficially like individual values are, in essential part, constituted on the social level.

The general moral of this chapter should now be evident. Rational choice theory has provided some profound illumination of interesting processes of decision-making and of interactions between human decision-makers. Economics, the theoretical enterprise that has most systematically developed the perspective of rational choice theory, has the potential for illuminating processes of production and distribution of goods, etc. But the attempt to present either the general or the more specific version of this perspective as a general account of human behaviour is futile. By itself rational choice theory assumes too narrow a view of the springs of human behaviour, too simplistic a view of the relation between the individual and the social, and too scientistic a view of the relation between facts and values. In conjunction with more explicitly social-level investigations of the framework in which individuals operate, and more directly interpretative approaches to unravelling the systems of meanings societies construct at the social level, rational choice theory may prove to have an important role among the set of tools that we bring to understanding human behaviour. But it can be no more than one among a range of such tools.

7

Freedom of the Will

1. Introduction

So far this book has mainly been about science, or some pretenders to the title of science. It may therefore seem surprising that I choose to conclude the book with a discussion of one of the most ancient and traditional problems of pure philosophy, freedom of the will. I hope the surprise will be short-lived, however. It is the sciences of the human mind, and their aspirations to provide complete explanations for human behaviour, that have naturally focused recent worries about the intelligibility of free action. And, often no doubt at a somewhat inchoate level, it is concerns about the threat to human autonomy that have motivated hostility to some of these scientific, or scientistic, projects. My own view is that the connection between these issues is more indirect. It is the metaphysics that underlies and motivates scientism that also grounds doubts about the possibility of human autonomy. In other words, the sciences that seem superficially inimical to human freedom are based on a metaphysics that really does preclude freedom. My aim in this chapter is to show how disposing of both the bad metaphysics and its scientistic spawn opens the way for a proper account of human autonomy. Such an account finally is one of the vital ingredients we need for a more complete understanding of humans in the societies that are their natural and necessary environments.

The argument of this book so far has been that a proper understanding of a domain as complex and richly connected to diverse factors as that of human behaviour can only be adequately approached from a variety of perspectives. But it is not at all obvious how concerns about human autonomy are defused by the move from a single reductive approach to explanation to a pluralistic approach. In this chapter I try to show how the pluralistic framework I have defended in previous chapters does indeed leave space for real human autonomy. I begin by defending the thesis, in sympathy with most untutored convictions but contrary to the orthodox philosophical view of the subject, that the denial of determinism does remove some of the central difficulties in providing a proper account of human freedom. I shall also briefly explain what I take to be some of the grounds and consequences of the denial of determinism.[1]

2. Free Will and Determinism

It has notoriously been supposed that the doctrine of determinism conflicts with the belief in human freedom. Yet it is not readily apparent how indeterminism, the denial of determinism, makes human freedom any less problematic. It has sometimes been suggested that the arrival of quantum mechanics should immediately have solved the problem of free will and determinism. It was proposed, perhaps more often by scientists than by philosophers, that the brain would need only to be fitted with a device for amplifying indeterministic quantum phenomena for the bogey of determinism to be defeated. Acts of free will could then be those that were initiated by such indeterministic nudges. Recently there has been some inclination to revive such a story as part of the fallout from the trend for chaos theory. Chaotic systems in the brain, being indefinitely sensitive to the precise details of initial conditions, seem to provide fine candidates for the hypothetical amplifiers of quantum events.[2]

But this whole idea is hopeless, and appeals to quantum mechanics merely illustrate the hopelessness.[3] To see this, one needs only

---

[1] This task is undertaken in greater detail in Dupré, 1993a: chs. 8 and 9.

[2] More recent appeals to exotic physics for the solution of this problem have involved Francis Everitt's many-worlds interpretation of quantum mechanics. On one reading of this, the world is constantly splitting into two, and one cause of some such splittings might be free acts of will. On what appears to be a more sophisticated version of the theory there is only one world, but it consists of a vast superposition of coexisting states. One philosopher who thinks this view helps us to understand the mind is David Chalmers (1996). Although Chalmers does not directly address questions of freedom, he presents the individual mind as simply one among all the possible paths through the superpositions of minds that each mind is at any moment part of (or so I understand him; see especially his p. 353). This view, I suppose, explains the illusion of freedom as made possible by a certain kind of randomness. My own view is that this is one of the several points in Chalmers's book where he has failed to draw the conclusion of an argument *modus tollens*.

[3] No doubt the belief in the existence of indeterministic events at the root of free action was often also connected with the inchoate hope that these might be sufficiently loose and microscopic that even an immaterial soul might have a chance of subtly influ-

to consider that the interest in establishing free will is not the conviction that humans are random action generators, but a concern
that human autonomy is inconsistent with the possibility of fully
explaining human actions in terms that have no apparent connection
with the wishes and beliefs of the human agent. Standard compatibilist claims that human autonomy and mechanistic causal
explanation are not mutually exclusive may or may not be defensible.
But the attempt to reconcile human autonomy with the complete
randomness of human actions is surely a dead end.[4] At first sight it
appears that, despite the initial worries about determinism, indeterminism makes the conception of freedom of the will even less
supportable.

A great deal of recent discussion of these issues has concerned the
question whether, in saying an action was free, we imply that the
agent could have done otherwise. It is generally assumed that if
determinism is true, then the agent could not have done otherwise
and therefore that, if the implication just mentioned holds, we cannot be free. The problem that will be clear from the preceding discussion is that even if determinism is false, it is far from obvious
what could make it true that the agent could have done otherwise in
a way that does anything to illuminate any doctrine of free action. In
particular, we had better not interpret it as meaning just that the
action was produced only with probability, not certainty (this may be
true; it just doesn't add anything relevant to freedom). At any rate, I
am not much concerned here with the question of whether the agent
could have acted otherwise—my inclination, in fact, is to think this
is largely irrelevant to questions of freedom. My concern is rather
with autonomy, in the sense that the agent can be seen as truly the
originator of an action. This idea can be found most clearly in the
concept of agent causation, deriving from Roderick Chisholm
(1964). This is the concept of an agent as an initial source, rather
than merely an intermediate link in the chain, of causality—or as he
says, a 'prime mover'. For Chisholm it is an idea directly motivated by
the thought that a responsible agent must have the power either to
act or to refrain from acting. Chisholm also seems to think that this

---

encing them. Though this makes the idea less absurd from the point of view of understanding human autonomy, it introduces new absurdities that I cannot attempt to address
here.

[4] This point was clearly stated by C. D. Broad (1952).

is a kind of causality unique to agents. I differ from Chisholm, first, in doubting the importance of alternative possibilities, and second, in that I take agent causation to be much more similar to causation in the non-human realms than he allows.

My aim in this chapter, then, will be to show that the solution to the problem of the freedom of the will does lie, despite familiar objections, with the truth of indeterminism. But the point of this demonstration will not be to defend the idea of alternate possibilities (though I am, in the end, sympathetic to this), but to provide a proper account of autonomy in something like the sense of agent causation. The first step in this account is to distinguish two very different grades of indeterminism. The indeterminism entailed by the common understanding of quantum mechanics, while it denies that the causal upshot of a situation is a determinate function of any fact about that situation, still insists that there is a complete causal truth about every situation. It is just that this truth is in the form, not of a unique outcome, but of a range of outcomes with specific probabilities attached to their occurrence. Thus situations are still conceived as evolving according to laws, just laws of a somewhat different kind. I shall refer to both traditional determinism, and this brand of moderate indeterminism, as versions of the thesis of *causal completeness*. Even if determinism is false, causal completeness requires that there be some quantitatively precise law governing the development of every situation. If we maintain the doctrine of causal completeness, then the only retreat from physical determination of our actions is in the direction of more or less unreliability, hardly a desirable philosophical goal. However, the indeterminism that I wish to advocate is something quite different, the denial of causal completeness.[5] I shall maintain that few, if any, situations have a complete causal truth to be told about them. Causal regularity is a much rarer feature of the world than is generally supposed. And the real solution to the problem of freedom of the will, I shall argue, is to recognize that humans, far from being putative exceptions to an otherwise seamless web of causal connection, are in fact dense concentrations of causal power in a world where this is in short supply.

The solution to the problem of human autonomy that I propose, then, is a complete reversal of traditional non-compatibilist

---

[5] Causal completeness remains the orthodox assumption in the philosophy of science. It has been criticized in most detail by Cartwright (1999).

approaches. Such solutions have assumed that the non-human world consists of a network of causal connections, the links in which instantiate lawlike, exceptionless generalizations, but have tried to show that humans, somehow, lie outside or partially outside this web.[6] By contrast, I am suggesting that causal order is everywhere partial and incomplete. There is no such causal web. But humans, by virtue of their enormously complex but highly ordered internal structure, provide oases of order and predictability; they are potent sources of causality. Thus the significance of recognizing indeterminism is not at all to show that human actions are unreliable or random. It is rather to show that the causal structure that impinges on human beings, whether externally from macroscopic causal interaction, or internally from constitutive microstructural processes, is not such as to threaten the natural intuition that humans are, sometimes, causally efficacious in the world around them.

This picture immediately accords with some obvious empirical facts: among the most apparently orderly features of the external world, such as straight roads and vertically stable edifices, not to mention complex machines, are products of human action; and among the most predictable entities in the world, as Hume, to a rather different purpose, argued, are people. Plans can be coordinated among many people, and complex human institutions can function, because human behaviour is to a substantial degree reliable; as Hume remarked, purses of gold don't remain undisturbed for long at Charing Cross. All of this is quite unproblematic if we see humans as sources of causal order rather than either as exceptions to a universal external order or as insignificant components of some all-encompassing cosmic order. Thus a radical rejection of the traditional mechanistic assumption of causal completeness does indeed do something to defuse the traditional problem of free will.

I shall expand on these claims in the course of this chapter. Prior to that, however, my main task will be to render its presuppositions plausible. The orthodox, though certainly not the universal, contemporary view of free will is compatibilist: it holds that everything we have any right to want from freedom of the will can be had in a deterministic world. In the next part of the chapter I shall argue, on the other hand, that determinism, specifically microphysical

---

[6] A classic statement of such a position is that of William James (1884/1956).

determinism, or, in fact, merely microphysical causal completeness, really is a problem for an adequate account of human autonomy. But, as I argue in the succeeding section, we have, fortunately, no reason to believe in determinism—or even causal completeness, whether microphysical or of any other kind. The chapter will conclude with some further discussion of how I conceive the rejection of causal completeness to provide a way out of the traditional problem of free will.

3. Microphysical Determinism and the Causal Inefficacy of Everything Else

Suppose that there is some set of microscopic entities undecomposable into any smaller constituents, and of which all larger entities are composed. Assume that all putative entities that might appear not to be composed of anything (numbers, abstract objects, universals, etc.) are either wholly dependent for their existence and behaviour on objects made of these microscopic entities, or nonexistent. Though these suppositions are hardly uncontroversial, I believe that they would be widely accepted among the many philosophers who think of themselves as physicalists. Now suppose that we also have a fully deterministic account of the behaviour of these microscopic entities. Although heroic attempts have sometimes been made to deny it, it seems to follow inevitably from this set of assumptions that the behaviour of everything is fully determined by the laws at the microlevel. This seems to follow immediately from the assumption that objects at higher levels are composed entirely and exhaustively of the microscopic objects. For, given the assumption of determinism, it is true of every individual microscopic object that its behaviour is fully determined by the laws governing microscopic objects. And surely if the behaviour of every constituent of a thing is determined, so is the behaviour of that thing.

This point can be made more graphic by thinking of a constituent of a human being, say an electron in my finger. I might be inclined to explain the movement of that electron by saying, for example, that I was waving to a friend, and my waved hand just brought the electron along with it. But clearly this explanation is going to have to be consistent, at the very least, with the explanation in terms of the microphysical laws acting on the electron. If we now consider the same condition applying to all the various electrons and suchlike in my arm, it would appear that only cosmic coincidence or some kind of dependence of the higher level on the processes at the lower level could ensure this overall compatibility. The bold conclude at this point that either the higher-level phenomena are reducible, in the sense of derivable, from the lower-level phenomena, or that the former cannot really exist at all (eliminativism). The more cautious fall back on claims of supervenience. They claim, that is, that even if the connections between the lower and higher levels are far too complex for us to discern any systematic relationships, the latter do, nonetheless, depend entirely, or *supervene*, on the former. As far as I can tell, however, this is merely reductionism with a modest reticence about the capacity of humans to carry it out. At any rate, none of these positions allows any genuine autonomy to the higher structural level.[7]

One might wonder whether, even given that the laws at the microlevel fully determine the physical trajectory of my arm, they might nevertheless fail to determine that that movement constitutes waving to a friend. So the microphysical can explain why my arm waved about, but fails to say what I was doing. But if this is right, it can only mean that we need to push the microphysical explanation further back. Facts such as that my friend was leaving and that I thought it polite to wave to him and wanted to be polite, etc., etc., cannot, on the picture under consideration, deflect the electrons from their predetermined paths. If such facts, the facts that make it true that what I was doing was an act of waving to a friend, have a role to play in the explanation, it can only be through their correspondence with, perhaps even identity to, the underlying microscopic facts. So perhaps to capture what we might otherwise have supposed to be the correct, mentalistic, explanation we need to bring in more microphysical facts than we might if we were only looking for microphysical antecedent conditions that cause my arm to move. But nothing has suggested, or could suggest, that the resources required could lie outside the realm of microphysics.

---

[7] This is at the basis of Kim's well-known arguments against non-reductive physicalism (Kim, 1993, especially essays 14 and 17). Kim shows that such a position requires 'downward causation', the causal influence of macroscopic on microscopic entities. I accept the argument but, as will be clear below and as I have explained elsewhere (Dupré, 1993a), I see no problem with downward causation.

It is important to stress the concept of causal completeness rather than merely determinism here, since nothing is significantly altered in the preceding argument by moving from a deterministic to an indeterministic but complete system of laws at the microlevel. Given my intention to drink from the glass of water in front of me, the probability that the electron referred to two paragraphs ago will move in a certain direction is very high. Again there must be some parallel explanation at the microlevel that also attributes a similar high probability to such a move. And again, when we aggregate all the particles that compose my arm, some explanation is required of the apparently extraordinary coincidence between the phenomena at the two levels.

The compatibilist will not, perhaps, be particularly disturbed by any of this. My action, according to the compatibilist, is caused by my beliefs and desires or whatever internal states, and these internal states are also physical states of my brain. She will invite us to recall that autonomy *requires* that our states of mind should be causally efficacious, and to agree that to be so they must be part of the causal nexus of the physical. I have just two comments to make to such a compatibilist. First, if the arguments that have driven so many physicalists away from robust reductionism towards doctrines of supervenience, anomaly,[8] or even eliminativism are correct, the supposedly causally efficacious mental states appear to be in a precarious state. If eliminativism were true, they would be in the worse than precarious state of nonexistence. But even on other such putatively non-reductive physicalist doctrines, there will be no determinable principles on the basis of which a physical causal process will give rise to the causal processes at the mental level, and the co-occurrence of processes at these levels will be something of a mystery.

But the second and more important point is that even if it is clear how a physical process of the kind occurring must at the same time give rise to mental processes or events of the appropriate kind, it is impossible to escape the charge that these processes are redundant. All the physical movements of the agent would have happened even if the mental occurrences had not, if, that is to say, there had been no principles or laws requiring mental processes or events to come along for the ride with the physical ones. The mental is at best

---

[8] I refer, of course, to the anomalous monism of Davidson (1970).

epiphenomenal,[9] which is to say lacking in any autonomous causal power. In summary, then, causal completeness at the microlevel must entail reductionism, at the very least in the sense of the supervenience of everything else on the microphysical. And even supervenience, I claim, is sufficient to deny any real causal autonomy to higher structural levels.

The alternative picture I would like to advocate denies causal completeness at any level. Objects at many, probably all, levels of the structural hierarchy have causal powers. One of the reasons why these causal powers are never displayed in universal laws (deterministic or probabilistic) is that objects at other levels often interfere with the characteristic exercise of these powers.[10] I take it that the example of the electron in my hand is best seen as such a case, a case, that is, of interference with the microphysical level by the macrophysical.[11] If that is right, then the behaviour of microlevel objects is very frequently consequential on processes at higher structural levels. As a simple example in the opposite direction, a person's plans can be seriously impeded by a dose of radiation. Just as I advocate ontological pluralism at particular structural levels against the essentialism that tries to insist on a uniquely privileged position for one set of kinds, I want to insist that the same ontological tolerance should be accorded between structural levels. As objects are united into integrated wholes they acquire new causal properties (perhaps that is exactly what it is for a whole to be—more or less—integrated). I see no reason why these higher-level wholes should not have causal properties just as real as those of the lower-level wholes

---

[9] The recognition of this consequence of the completeness of physics has led to some extraordinary discussion of consciousness, the thing that seems most clearly to be shown thereby to be epiphenomenal (see Chalmers, 1996).

[10] As of course may objects at the same level. This kind of objection to universal regularities has been emphasized by Cartwright (1983), who has developed the idea in terms of the inevitability of open-ended *ceteris paribus* conditions.

[11] It will be objected—and may already have been objected—that the electron will be pushed by the microscopic object or objects immediately behind it and will push those in front of it, and thus all the particles are moving in response to microlevel forces. I do not mean to deny this: certainly it would be absurd to suppose that my intention independently acted on each particle in my arm. The real issue is whether all these arm-particles are moving as part of a much wider set of microphysical events (photons bouncing of the glass, hitting my retina, stimulating my brain, etc.) of which my intention to drink the water is ultimately a mere epiphenomenon, or whether, rather, the fundamental explanation for all those particles pushing one another in a certain direction is that I am thirsty and see a glass of water I plan to drink. Evidently I prefer the latter view.

out of which they are constructed. The reasons that many philosophers have seen for denying such higher-level reality are all
grounded in the conceptual nexus that links determinism (or at least
causal completeness) and reductionism. To the former, which is both
still widely believed and traditionally linked to concerns about
human freedom, I shall now turn.[12]

4. Causal Incompleteness

The thesis I shall now defend is that there is no plausible ground for
the belief in causal completeness. I shall address most of the argument to the doctrine of determinism, but I intend that everything I
say will apply equally to indeterministic versions of causal completeness unless I explicitly differentiate the two cases. The basic strategy
of my argument will be as follows. Presumably determinism is a very
strong metaphysical assumption. To claim that everything that happened had to happen, given the totality of prior conditions, is to
impose an enormously strong—indeed the strongest possible—
restriction on the possible evolution of the universe. And even the
claim that the state of the universe at any time fully determines a set
of objective probabilities for its subsequent state is a strong assumption. My point is then that such strong assumptions require persuasive reasons if they are to have any plausibility. I do not take
seriously the idea that determinism might be established by means of
a transcendental argument of some kind, simply because, as I shall
explain below, an indeterministic, causally incomplete world seems
to me entirely possible. Thus the first step in the argument is to insist
that the onus of proof belongs with the determinist. Determinism
has been so closely linked with the philosophy underlying the rise of
modern science that it has come to seem obvious, something to deny
which is to call in question the whole scientific project. But in my
view determinism is a philosophical free rider on the scientific world
view, something for which the latter provides no warrant, and something that is, despite the success of the scientific world view, in-
herently implausible. Assuming that what is most fundamental to the
scientific world view is empiricism—very broadly, answerability of

[12] The various theses summarized in this paragraph are defended in detail in Dupré
(1993a).

our beliefs to what we actually experience in the world—my question will be whether there is any basis in our empirical interaction with the world for supposing that it is causally complete. My answer will be in the negative.

There are two main kinds of experience that might be held to legitimate a belief in determinism. These are, first, our familiarity with scientific laws and, second, our everyday causal experience. An important special case of the latter is our experience of highly organized systems, especially machines and organisms. I shall deal with these topics in turn, but reserving the special cases of machines and organisms to a separate section. First, then, do the results of scientific investigation lend support to the idea that the world is deterministic, or at any rate causally complete? Here I must first dispose of a troublesome red herring. It is often claimed that science must assume determinism as a methodological imperative. The idea is that it would be sheer defeatism, when confronted with a phenomenon anomalous in the light of current belief, to assume that this was simply a phenomenon outside the causal nexus. We naturally and correctly attempt to broaden our understanding of the range of phenomena in question so as to remove the appearance of anomaly. But that, it is claimed, is to assume that the anomalous phenomenon is in fact part of a uniform and complete causal nexus. Thus it might be suggested (and this is the non sequitur) that science must assume determinism; and then, perhaps, that the successes of science provide evidence that the presuppositions of science, in particular determinism, must be true. But of course to say that science aims to explain phenomena does not entail that all phenomena can be fully explained. And to say that science has had some explanatory successes hardly implies that everything that happens can be fully explained as part of an underlying universal regularity. We can be optimistic about life without inferring that this must be the best of all possible worlds.

So do the actual results of scientific research provide more direct evidence for determinism? The most compelling such results, for the reasons spelled out in the preceding section, would be those that provided evidence for causal completeness at the microlevel. But clearly there is no such evidence. Although certain very specialized phenomena in extremely carefully controlled conditions do exhibit some impressive regularities, this is the entire extent of such evidence. (As should become apparent later on, the fact that these regularities are produced in extremely elaborate *machines*—machines painstakingly designed for the very purpose of producing these regularities—is of great significance.) Evidence for causal completeness would require that increasingly complex systems of physical particles could be shown to be amenable to causal explanation in terms of the laws said to govern individual particles, evidence, that is to say, for general reductionism. I cannot here go into the general difficulties that confront the project of reductionism. But I do not need to do so. No one has claimed to be able to explain the behaviour even of very small collections of particles in terms of the behaviour of individual particles; the reduction even of relatively simple parts of chemistry to physics is now looked on with considerable scepticism (Scerri, 1991; 1994); and even physics itself is acknowledged to consist of laws the relations between which are obscure, though at least the unification of physics is still looked upon by some physicists as an attainable goal. At any rate, the view that every physical particle has its behaviour fully determined by microphysical laws must derive any plausibility it has from some source other than the development of microphysics.[13]

It appears then that microphysical determinism must be motivated, somewhat paradoxically in view of the connections between determinism and reductionism, by experience at the macroscopic level. But before turning to our everyday experience of causal regularities we might consider the possibility that microphysical determinism could be motivated by our knowledge of macrophysical laws. The obvious candidates, since they remain the most widely admired paradigm of scientific knowledge, would be the laws of Newtonian mechanics. But here we encounter exactly the same difficulty that we saw at the microlevel. Whereas scientists have been able to subsume very simple systems such as the solar system under impressively reliable regularities, the ability to apply Newtonian laws to more complex systems has proved severely limited. The notorious failure to solve the three-body problem, let alone n-body problem,

[13] It is of course true that microphysical laws *purport* to apply to indefinitely complex systems, in the sense that they determine how the formalism should, in principle, be applied to such systems. But in practice they certainly cannot be so applied. And one need hardly be a radical sceptic about induction to resist extrapolation from a very narrow and limited set of data to every phenomenon whatever that could in principle be subsumed under the purported regularity. For a detailed and subtle account of how results really are generated in physics experiments, see Galison (1987).

marks this failure. Thus we have no empirical evidence for the general truth of Newtonian mechanics as applied to complex systems of bodies unless we are prepared to countenance inductions grounded on one kind of case (very simple systems) to all cases, most of which are very different from those empirically studied. Moreover, to reiterate a point emphasized by Cartwright (1983), we know that laws such as those of Newtonian mechanics are true only under a very stringent *ceteris paribus* condition, a condition we know to be generally false. Thus, far from knowing that these laws are universally true, we know that they are generally false. The assumption that the laws of Newtonian mechanics are, in some sense, carrying on regardless under the overlay of increasingly many interfering and counteracting forces is not merely sheer speculation, but actually of dubious intelligibility. What are these laws supposed to be doing, given that the objects, subject to such diverse other influences, are not behaving in any sense in accord with them? Certainly this can hardly be a good *empirical* ground for the alleged universality of microphysical laws.[14]

The other common idea, mentioned above, is that determinism is evident from our everyday experience of causality. This assumption can be seen in classical regularity theories of causality from David Hume to J. S. Mill and J. L. Mackie.[15] Hume appeared to take determinism outside the human sphere to be so obvious as not to need much discussion. He was more concerned to show, with well-known examples such as the sure and swift appropriation of a purse of gold abandoned at Charing Cross, that humans were subject to regularities just as immutable as those governing the natural world. Mill was a good deal more sensitive to the complexities of regularities of the latter kind, realizing that the regularities of common experience could easily enough be defeated by either the absence of necessary background or auxiliary conditions, or by the presence of interfering conditions. Thus a lighted match thrown onto a pile of dry straw will always start a fire—unless, that is, there is no oxygen, or a fire extinguisher is simultaneously directed at the straw, etc. While thus acknowledging the complexity of everyday causal regularities, Mill appears to have thought that with sufficient care to include all the relevant auxiliary conditions and exclude all possible

---

[14] See Suppes (1993) for a more detailed argument complementary to the present one.
[15] See principally Hume (1748), Mill (1875), and Mackie (1974).

blocking conditions, a truly universal regularity could be discovered. This idea reached its most sophisticated expression with Mackie's analysis of an everyday cause as an insufficient but non-redundant part of an unnecessary but sufficient condition, or an 'inus' condition. The sufficient condition in this analysis is the cause with all the auxiliary conditions and the negation of possible interfering conditions. The non-necessity of such conditions points to Mackie's additional recognition that there might be many such complex sufficient conditions of which none, therefore, would be necessary (a bolt of lightning might equally well have ignited the pile of straw).

Many objections can be raised against this picture, at least if it is assumed that it intends one to take seriously the universality of the implied laws rather than merely to illuminate the relations between miscellaneous items of causal lore. One may well doubt, to begin with, whether there is any definite limit beyond human imagination to the number of conditions that we might need to add to produce a fully universal generalization. More seriously, the more conditions are added, the further these putative regularities recede from any possibility of empirical support or refutation. Indeed the reason we are forced to move from simple regularities (e.g. lighted matches cause fires in flammable materials) to increasingly complex and qualified regularities is simply because we recognize the general falsity of the simpler ones. But as we move to such ever more complex regularities, first, the amount of evidence even bearing on the truth of the regularity will rapidly decline; and second, in keeping with the process that brought us the complex regularity in the first place, were we to find an exception to the complex regularity we would presumably respond by looking for a further interfering condition rather than by rejection of the entire regularity. This suggests that the Mill/Mackie programme might better be seen as embodying a methodological rather than a metaphysical conception of determinism.

A second kind of objection casts doubt on the empirical basis of everyday causal determinism from a rather different perspective. Many everyday phenomena give no superficial appearance of being deterministic or even nearly deterministic. Consider, for example, a tossed coin. Now it is often asserted that a coin spinning through the air is a fundamentally deterministic phenomenon, and the only reason we are unable to predict the outcome is that we have an insufficiently precise knowledge of the initial conditions. It is much less clear why this is asserted. Presumably it must be because the kinds of laws involved in such a process (mainly Newtonian) are assumed to be deterministic. But I have already considered the weakness of that line of thought. The present case, since it is one in which we cannot in fact make any such predictions, provides further support for the argument against basing determinism on macroscopic scientific laws. At any rate, the thesis that everyday causal experience, suitably refined in the style of Mill and Mackie, provides grounds for the belief in determinism, simply ignores the fact that a great deal of our experience, whether of gambling devices such as tossed coins and roulette wheels, or just of seemingly quite erratic natural phenomena such as falling leaves or swirling smoke, provides no such grounds.

The final argument I shall mention is perhaps the most telling. It is that if there is causal indeterminism anywhere, it will surely be (almost) everywhere. Suppose, as is sometimes rather bizarrely suggested, that the only locus of indeterminism is in quantum mechanics. But surely—and here phenomena such as hypothetical quantum amplifiers in the brain have genuine significance—it must be impossible to insulate the indeterminacy of quantum events so fully from consequences at the macroscopic level. Consider again, for instance, the tossed coin, and suppose that its trajectory deterministically produces—*ceteris paribus*—its final outcome. Suppose the coin is at a point at which it is about to land heads. And suppose finally that a collision with a fast-moving air molecule is sufficient to reverse this outcome and produce a toss of tails. If the situation is sufficiently delicately balanced this must surely be possible. Assuming that the molecular trajectory is a sufficiently microscopic event to be subject to some degree of quantum indeterminacy, then we can easily see that the claim to determinacy of the coin-tossing event cannot be sustained. We cannot treat this as merely another interfering factor, because whether or not it has any effect on the final outcome cannot be determined by any amount of knowledge of the initial conditions.

It is a further advantage of this example that a coin toss is the kind of event that might imaginably have massively ramifying consequences. Perhaps the last degenerate scion of some aristocratic line is wagering his fortune on this coin toss. The outcome will dramatically affect the lives of his dependants, servants, creditors, etc. and their fortunes will have an increasing cascade of consequences. The general point that this example is intended to illustrate is that indeterminism anywhere, by virtue of the variety of causal chains that might be initiated by an indeterministic event, is liable to infect putatively deterministic phenomena anywhere. It is significant that this applies equally within and across levels of structural complexity.

One final point will conclude this part of the discussion. The last argument presented is an argument against determinism, but not necessarily against causal completeness. In the case of the coin toss, provided only there is no correlation between interfering molecular events and outcomes, we should expect that these would be equally likely to change heads to tails and vice versa. So even if these interfering events occurred in accordance with no law even of a statistical nature, they might not render incomplete the supposed law that coins of a certain kind come up heads 50 per cent of the time. On the other hand the preceding arguments, based ultimately on the lack of empirical support for determinism, seem if anything even more pressing against an indeterministic version of causal completeness. For any investigation of a range of phenomena will provide statistical facts. That, for some x, x per cent of events of type A are followed by an event of type B, is a matter of logic. But for this very reason, even if we have excellent grounds for believing that As really do have a tendency to produce Bs, it is difficult to see why we should be led to believe that there is any x such that it is a *law* that x per cent of As produce (or are followed by) Bs. The most plausible basis for such a belief, I suppose, would be microphysical reductionism, a topic about which I shall say no more here. We might better ask, 'What would it *mean* for there to be a law of this kind, as opposed to there merely being a tendency of As to produce Bs, and a statistical correlation of a certain strength between As and subsequent Bs?' Ignoring for the present purposes a range of widely explored subtleties concerning spurious and genuine correlations, joint effects of a common cause, and so on, which would be required for a detailed answer to this question, the simple answer which is sufficient for my present purposes is just that a precise causal law should license us to expect that the proportions measured in a suitably large number of trials should be (approximately) repeated in the future. It seems to me, on the contrary, that in practice such an expectation would often be foolhardy.

In real life, the level of confidence with which we treat statistical experience as a guide to future expectations will vary from almost total to almost none. No doubt many explanations could be given of the reasonableness of such a perspective, some consistent with causal completeness. The explanation that seems to me most consistent with both investigative practice and the experience of causal regularity, however, has nothing to do with laws or statistical uniformities at all. Correlations reveal, I believe (subject to well-known qualifications), the causal powers of certain objects or events to produce particular effects. Whether we expect the production of such effects to occur with a fairly constant frequency depends whether we think that the frequency of other relevant causal factors is likely to remain reasonably stable. But without some apparently quite arbitrary way of privileging a particular constellation of background conditions, there is no such thing as the quantitatively precise, constant, and timeless tendency of As to produce Bs *ceteris paribus*. Other things can be a particular way, and they can be more or less reliably that way. But except in the very simplest cases, as in Newtonian mechanics where we imagine that there are only two bodies in the universe, and everything else is supposed not equal but absent, I do not know what 'everything else being equal' even means. Thus once we have fully appreciated the complexity of the causal nexus, the thesis of indeterministic causal completeness is seen not only to be devoid of empirical support, but even to be, once again, of dubious intelligibility.

5. Machines and Organisms

As I have tried to show in the preceding section, I do not think that direct reflection on our (extremely limited) knowledge of universal regularities lends much support to the idea of a universe with a complete causal structure. However, it may well be that deterministic intuitions derive more from reflection on complex and highly organized structures, especially machines and biological organisms. Since the overall metaphysical vision out of which the whole problem of free will arose is aptly referred to as mechanism, it is certainly appropriate to consider the artefacts that have somehow come to provide a model for the universe; the consideration of organisms, notoriously liable to be treated as a kind of naturally occurring machine, will bring us back to a topic fundamental to the central theme of this book, the causal status of humans.

It is easy enough to see why machines should have some tendency to inspire deterministic intuitions. Machines, good ones anyhow, are extremely predictable. I am confident that the text I type into my computer is exactly what will eventually come out of my printer when I connect them up in the right way. (Though not so confident that I do not occasionally make a hard copy; and some people, I am told, even make back-ups of their computer files on disks.) But a little further reflection makes it very puzzling that something like this, rightly admired as one of the great triumphs of modern technology, should be taken as a model for the universe in general. If the sort of regularity that is characteristic of a good computer or car were typical of the universe it would, one might imagine, be fairly easy to make, or perhaps even just find, such things. But it is not at all easy, which is why such technological achievements are admired. If the universe is a machine, it is far from obviously so.

Perhaps a more sympathetic interpretation of the tendency for machines to inspire determinism is the idea that only if determinism were true would it be possible to make reliable machines. And since we can make reliable machines, determinism is proven to be true. Underlying what seems to me a great exaggeration in the first premise of this argument there is, nevertheless, a very interesting question: what degree of order must exist in the world for the kinds of reliable machines we possess to be possible? The beginning of a more temperate answer to this question than the immediate appeal to determinism is the observation that no machines are completely reliable, and some are very unreliable. The point of this observation is not to insist—though strictly speaking it is no doubt true—that there is some possibility, however remote, that when I type the word 'type' on my computer a four-letter obscenity will instead appear on the screen; or that when the spark ignites in the combustion chamber of my car the gasoline inside it will spontaneously liquefy. Rather I want to focus on the question, what it is that makes machines more or less reliable. And of course the answer is not, at any rate, that reliable machines have access to more universal laws.

Consider, then, what is by modern standards a fairly simple machine, an internal combustion engine. If we ask how such a machine operates we may be content with a very simple story: a mixture of air and petrol is exploded in a cylinder, pushing a piston down the cylinder; the cylinder is connected to a shaft which is rotated by the moving piston. A number of similar cylinders are connected to this shaft, and a sequence of explosions keeps the shaft rotating continuously. It seems to me that this is, roughly speaking, a correct answer to the question how an internal combustion engine works. But if, on the basis of this explanation, someone lined up some coffee cans partially filled with petrol on the kitchen floor, stuck toilet plungers in the cans, tied the ends of the plungers to a broomstick, and then posted lighted matches through little holes in the sides of the coffee cans, they would certainly not have built an internal combustion engine (though I suppose the broomstick might jump about a bit).

I suggest that it is useful to think of how a machine works in two stages. First there is the question what makes it even possible for the machine to do what it is supposed to do. A slightly more elaborate version of the answer sketched in the previous paragraph might be an answer to this question for an internal combustion engine. Having got that far, however, most of the details of the internal combustion engine concern the more or less ingenious auxiliary devices that make sure it really does do what it is intended to do rather than one of the many other things it has an initial capacity to do. So, for instance, the cylinder must be strong enough to avoid simply disintegrating when the petrol explodes; the crankshaft must be extremely strong and rigid if it is to reliably convert the linear momentum of the pistons to rotational motion; piston-rings prevent the energy of the explosion from being dissipated between the piston and the cylinder; oil must be provided to prevent the cylinders getting so hot as to seize in the cylinder, or for that matter melt; some way must be found to dissipate excess heat from the running engine; and so on. Even a Trabant has the capacity to run and sometimes does so. The difference between this and a well-designed car is that the behaviour of the parts of the latter is so tightly constrained that it can do nothing but what it is designed to do—though eventually, of course, even the best-designed machine will break free of its constraints. My point so far is just that this kind of constraint is not something characteristic of nature generally, but something that engineers devote enormous efforts to attempting, never with total success, to achieve.

Of course, this account of the reliability of machines does assume the reliability of various causal relations: gasoline and air mixtures almost invariably explode when sparked; heat will flow from a hot engine to cooling water circulating over it; and many others. It is interesting that many such regularities can be seen as reflecting the overall upshot of very large numbers of similar though indeterministic processes at the microlevel, which suggests the hypothesis that it is just those macrolevel processes that can be roughly reduced in this way that reveal this near determinism. But I do not want to insist on this here. While machines could presumably not work without exploiting extremely reliable regularities such as those just mentioned, the regularities that characterize the machines themselves, as with many other macroscopic causal regularities are only more or less reliable. And in keeping with a general philosophical theme I endorse, it is best to think of these regularities as involving the reliable exercise of the capacities of things when properly triggered and unimpeded. The capacity of suitable mixtures of petrol and air to explode when ignited is an extremely reliable one and very difficult to impede. Such reliably exercised capacities are no doubt a precondition of the possibility of building reliable machines. But the existence of such capacities provides no basis at all for the conclusion that everything that happens is the exercise of a similarly reliable capacity. Indeed a great deal of experience—experience of the generally more or less unreliable and unpredictable natures of things—speaks against it.

Reflection on how good machines are engineered, far from making us think of mechanism as generally characteristic of the world, should make us realize how difficult it is to turn even little bits of the world into bits of mechanism. Though I have admitted that machines could not be made to work if there were not things in the world with capacities that, under certain circumstances, are exercised with (almost) complete regularity, it is important to note that these are quite different from the much more complex and much more tenuous capacities of machines. A fortiori, the things in the world are not limited to simple, reliable capacities of the first sort; and the things that happen in the world are not always, or even generally, the simple exercise of such reliable capacities.

Turning now to organisms, it is a familiar idea, especially following Descartes, that organisms just are machines. Natural theology until the late nineteenth century considered organisms quite explicitly as the products of a divine mechanic.[16] Mechanistic modes of investigation have had extraordinary successes in uncovering how metabolism, reproduction, and other basic biological processes

---

[16] The *locus classicus* is Paley (1802).

work. And even in the domain of behaviour, the complex but highly stereotyped performances of many insects in, for example, constructing and provisioning burrows for egg-laying have many of the characteristics of a well-designed machine. To the extent that the analogy is appropriate, the same remarks that I made about machines will apply to the relevance of organisms to the prevalence of causal regularity. Looking, however, at the other end of the organic scale, and most especially at humans, the parallel with machines has serious limitations.[17]

The fact that when, for example, I intend to walk down the garden path, my legs move in just the right way to maintain my balance and propel me forward is, I suppose, something that could be explained in a manner strongly analogous to the performance of a machine, though perhaps one more complex than any machine we have yet managed to construct. I suppose that the physiology and cell chemistry of muscle tissue explains how the physical movements are obtained, and a variety of sensory and neural mechanisms bring it about that the motion is steady and in the right direction, and that a vertical posture is maintained. Although this seems significantly analogous to the account I offered of the working of the internal combustion engine, we should now note that an internal combustion engine is in reality not a machine but a part of a machine. If we think now not just of an engine but of an entire car, an important class of features has yet to be mentioned. I am thinking of such things as the ignition key, the steering wheel, and the brake pedal, those devices by which the machine is made to act in a way conducive to the ends of its human operator. A reliable car, as opposed to a reliable engine, the latter of course being a necessary but insufficient component of the former, is one in which there is a reliable correlation between inputs to these controls and the behaviour of the whole machine. Thus machines are not sources of causal autonomy; they are, at most, instruments for furthering the causal autonomy of their users. The superficial, and I think also deep, disanalogy between humans and machines is that humans have no controls.

---

[17] I focus here only on what I take to be the extremes of the animal scale. I assume that higher mammals, birds, and perhaps higher molluscs, are more like humans than they are like the most machine-like of insects. But I shall make no attempt here to draw any more specific lines between different kinds of organisms, though this may be an important task for those concerned with our ethical responsibilities towards animals.

It may rightly be objected at this point that insects with simple stereotyped behaviours have no controls either, yet I have claimed that they are closely analogous to machines. There are two possible responses. First, a stereotypical performance might simply be produced in response to nothing at all. More typically and interestingly, a kind of behaviour might be triggered by some sensory input, the sense organs thus serving as devices for producing behaviour appropriate to the external circumstances. This is primarily what I have in mind in talking of the stereotypic and machine-like behaviour of certain insects: a certain stimulus triggers a sequence of behaviour. One might reasonably suggest that the sense organs in such an organism serve the functions of controls. There is, of course, a tradition of psychological investigation of humans that applies just this model to humans. Though in its crudest behaviourist versions it has been almost wholly rejected, the idea that sensory inputs, mediated by 'information-processing mechanisms', somehow elicit the appropriate 'emission' of behaviour is still widely, perhaps generally, assumed. This is a mechanistic model, though one in which the complexity of the machine is such that we as yet have no idea what it is designed to do in the innumerable situations it encounters.[18] Against this model, and as I have argued in more detail in earlier chapters, I propose that we should recognize that we were not designed at all, and consequently there is nothing we were designed to do in any situation.

Between two views that I have rejected, that we are random action generators and that we are machines, can be found the view that makes sense of human autonomy. Many parts of humans have just the characteristics of machines that I have emphasized in the preceding discussion, namely complex constraints that ensure the predictable exercise of some capacity of an organ or physiological system. But humans are fundamentally different from machines in that they have no controls. Self-control, in the sense of the absence of external controls, is of course nothing but the autonomy, or free will, that it was the goal of this chapter to illuminate. I have not attempted to refute the idea that sense organs might sometimes function as

---

[18] As should be clear from earlier chapters, evolutionary psychologists and their fellow-travellers do think we are designed, and do think they know exactly what we are designed to do, to survive and reproduce. I have said enough about what I see as the inadequacies of this unwitting attempt to appropriate the creationist tradition.

controls, in the sense that the input to sense organs might determine, via a complex intermediate causal chain, the behaviour of the whole organism. This is presumably roughly true of simple organisms. But it does not appear to be true of ourselves, except perhaps in purely reflexive actions, such as ducking to avoid a flying object. The reason we are so liable to think of ourselves in this machine-like way is because we are tempted by determinism. If the world is deterministic then my behaviour is causally necessary given the stimuli that impinge on me; and presumably the most important stimuli are sensory ones. The point of all the complex machine-like parts of me would then have to be just to make sure that the causally elicited behaviour was appropriate to the circumstances disclosed by my sense organs. And this is the concordance that, according to evolutionary psychology, millions of years of evolution have succeeded in bringing about. But the rejection of causal completeness allows a more natural view of things. My complexity of structure gives me a vast array of causal powers, a range of powers that would be inconceivable without that intricate machine-like internal structure. But the exercise of those powers, though obviously influenced by the circumstances I perceive myself to be in, ultimately depends on an autonomous decision-making process.[19] Once we see causal order as something special rather than something universal, there is no obstacle to seeing the human will as an autonomous source of such order.

One very important point should be added, which should ultimately profoundly modify the preceding point. For much of human behaviour, context is far more than a trigger that prompts the emission of behaviour. Human behaviour, recalling a philosophical truism, consists not merely of movements, but of actions, and social context plays a central part in determining what action is constituted by a particular movement. The action of signing a cheque could be no more than a meaningless wiggling of my hand without an extremely elaborate social context. Most important of all, what would otherwise be merely noise becomes articulate language only

---

[19] Philip Kitcher, commenting on this passage, raised the pertinent question whether the 'I' introduced in this discussion, if not fully comprehensible in terms of neurons, receptors, and so on, did not recapitulate old-fashioned dualism. For now I want only to insist that the 'I' refers to the whole organism, not just some neurologically salient bits of it. I shall try to make clearer below how I understand the 'autonomous decision-making process' that it undergoes.

because there is a society in which noises have meanings.[20] I have said earlier that the usefulness of reductive analysis is that whereas knowledge of the inner workings of things is typically quite insufficient to tell us what they will do, it is generally the way of explaining why they have the capacities to do the things they do. In the case of humans this is only true if we limit consideration to capacities described in the broadest way. The structure of our brains is no doubt such as to enable us to learn a language. But what we can say or do with that capacity is entirely determined by the social context that we have the good or bad fortune to find ourselves in. Although it has been the main purpose of this chapter to defend the idea that individual humans are potential sources of causal efficacy, it is with regard to capacities the possibility of which derives from the relation of an individual to a social context that this potential causal efficacy takes the most significant forms.[21] And as I argued in Chapter 2, distinctively human capacities are almost all dependent on a social context, or at least a relation between an individual and a social context. In the next section I shall try to make clearer how such capacities, the distinctively human capacities that are most thoroughly concealed by the scientistic and doggedly individualistic perspectives that have been criticized in this book, can help to make sense of human freedom.

6. Moral Autonomy

Pleasant though it would be to do so, I do not expect to resolve all the problems that have troubled philosophers over the ages concerning the human will. What I have, more modestly, been trying to show in this chapter is that, contrary to a notorious tradition of philosophical controversy, a reasonable metaphysics of causality presents no special difficulties for the idea of human autonomy, and requires neither ghostly nor random nudges of the physical causal order. In order to give more positive philosophical substance to the view, I shall begin this discussion by attempting to locate my views more perspicuously

[20] As I have mentioned earlier, the elaboration of this point is one of the most famous contributions of Wittgenstein's masterpiece, *Philosophical Investigations* (1953).

[21] Searle (1995, esp. ch. 6) provides an excellent though rather different account of the way social context shapes human capacities, though I doubt whether he would agree with the way I develop this thought.

within some aspects of the traditional debate, and in particular relate what I have said to some very famous views on the subject, those of Hume and of Kant.

Although it may well remain the dominant view of the subject, Hume's attempt to reconcile human autonomy with a classically deterministic structure of causal relations seems to me unconvincing.[22] On the other hand, Hume was surely right that exercises of human freedom were much better understood as instances of causality than of its complete absence. The intuition developed by Hume and subsequent compatibilists that a connection, probably causal, between the psychological states of the agent and the action is, far from a bar to autonomy, a necessary condition of autonomy, seems basically correct. But given the assumption that the agent is located in an otherwise seamless causal nexus, this insight closes off the only, acausal, escape from this nexus, and inevitably leaves the psychological causes of action at best supervening ineffectually above this self-sufficient nexus. His problem, therefore, was his commitment to a universalistic regularity theory of causality. Given such an account of causality, any departure from determinism is a failure of causality itself.[23] But the solution I have advocated was certainly not an option for Hume, since even more renowned than his compatibilist account of free will are his arguments against an ontology of causal powers; and it is just such an ontology that I present as one of the main resources for escaping the causal inefficacy of the human will. Although many philosophers remain convinced by Hume's arguments on this topic, an increasing number do not. This, however, is not a matter that can appropriately be addressed here, and I note only that I and others have attempted to rebut Hume's attack on causal powers at some length elsewhere.[24] Those who

[22] I cannot begin to discuss the enormous literature on this question. Strawson's classic paper (1974) perhaps brings out as clearly as possible some of the most troubling consequences of taking physical determinism fully seriously, though Strawson himself does not appear to think we should be led by such reflections to doubt the truth of physical determinism.

[23] Recently there has been a prominent movement to provide regularity theories of indeterministic causality (e.g. Eells, 1991). I believe there are deep internal problems with such a position (see Dupré and Cartwright, 1988; Dupré, 1993a: ch. 9). But at any rate, for reasons already explained, I do not think the move to such a theory will make any significant difference to the problems currently under consideration.

[24] Some recent advocates of causal powers include Harré and Madden (1975), Cartwright (1989), and Dupré (1993a: ch. 9). Anscombe's (1971) brilliant attack on Hume's account of causation is also highly relevant.

remain convinced that the idea of a causal power or capacity has been shown by Hume to be incoherent will not be convinced by this part of my argument.

Kant also appears to have thought that a deterministic causal structure was compatible with human autonomy, though the meta-physical excesses to which he was led in effecting this reconciliation have convinced almost no one. However, his conception of human autonomy offers some promise of providing a vital and final piece in the picture that I wish to present.[25] My point at the end of the last section was that human decision could be a real source of causal order in the world. However, this claim is liable to seem shallow without some further account of the origins of this order. In particular if one traces human decisions ultimately to contingent human desires, desires which presumably can themselves be traced either to our biological heritage or our upbringing, human autonomy seems at best a focus rather than a source of order. And this is just the con-ception of human decision-making discussed in the last chapter that has been cultivated for well over a century by economists and more recently by exponents of so-called rational choice theories in a var-iety of disciplines. It does not matter much at this point whether one says, with the economists, 'tastes are exogenous' (and, moreover, *de gustibus non disputandum*), or with the evolutionary psychologists, that they well up atavistically from our evolutionary past. Either way human autonomy turns out to amount to little more than the more or less effective attempts of a want-satisfying machine to satisfy its wants. Some account of the ultimate springs of behaviour that does more to explain the sense in which goals belong authentically to the agent who pursues them seems needed to give real interest to the account of human autonomy suggested in this chapter.

It seems to me that Kant's account of human action points in the right direction in which to look for this final ingredient of an account of human autonomy. Kant, as is well known, distinguished sharply between action motivated by desire and action motivated by principle. And for Kant only action motivated by duty, by the commitment to conform one's actions to the moral law, counted as truly free. Though this account has struck many as intolerably austere, and others as positively paradoxical in seeing free action as

---

[25] The ideas I borrow from Kant here are most accessibly presented in *The Groundwork of the Metaphysics of Morals* (1948).

law-governed,[26] it is easy to see how it addresses the concern of the last paragraph, that action merely directed at whatever a person happens to want seems to lack the authentic connection with the person acting that an account of free action would ideally incorporate. It does seem to me, however, that once the restrictions of the deterministic framework have been dispensed with, it is possible to gain this benefit of a very loosely Kantian account without either the extreme moralistic distinction in favour of action motivated by duty, or the metaphysically murky appeals to the noumenal world, which have combined to cast deep suspicion on Kant's conception. My suggestion is that it is quite generally the possibility of acting from a principle broader than the maximization of immediate satisfaction that grounds human autonomy. It is this, after all, that enables human action to produce order in the world, and it is this capacity above all that we aim to instil in our children through education.

In a world where order is a local and incomplete phenomenon, the importance of principle as a source of human action is easily stated: it explains how ideas, the creative acts of the human mind, can change the world. But unlike Kant, I do not make a fundamental distinction here between moral and more mundane principles, though perhaps it will be appropriate to make one of degree. I conceive of the principle 'Follow the architect's blueprints in determining where to build the wall' as as genuine a source of autonomous action as 'Do whatever is necessary to end hunger'. One does not have to be as severe in one's moral demands as Kant to see that doing whatever one feels like at the moment, if it is an intelligible human life at all, is not one that realizes the important human capacity for freedom. And despite this moderation of the Kantian position, the enormous importance of moral principles in this context should not be downplayed. The most fundamental reason why we should care about human autonomy is that it holds out the hope that human

---

[26] This again raises one of the main strands of debate that I have largely avoided during this foray into the free will problem, the question whether the agent could, when she acted, have done otherwise (see e.g. Frankfurt, 1969). For Kant, it is clear at least that if the agent had acted otherwise, she would not have acted freely. Being, up to a point, sympathetic with what I take to be the basic compatibilist insight, I am not convinced that this should be a fundamental issue. If it is, or should be, it is at any rate not one I have anything to say about here.

One theorist of free will whose views are in many respects highly compatible with my own is Eleonore Stump. Stump also argues against the view that a libertarian must insist on the possibility of the agent having acted differently (Stump, 1996).

action might produce a better world. And what that requires, I suppose, is action grounded in moral principles.[27] This is something I believe we are free to choose; and making this choice, I claim, can make a difference.

It is the curse of this topic that any suggestion of a basis on which autonomy might be grounded will inevitably provoke the question, 'What is the origin of that basis?' And hence what I have just said will surely invite the question, 'Where do principles themselves come from?' And the questioner is likely to perceive the standard dilemma: either principles are indeterministically embraced, reintroducing all the problems with naive indeterminism, or they are caused by the circumstances of the agent's upbringing and so on, thus again reducing the agent to just a complex part of the causal nexus. A better solution, I believe, lies with the point at which I ended the previous section of this chapter. Principles, I take it, are essentially linguistic phenomena. The ability to adopt a principle, and to make it part of one's nature that one aims to act in accordance with that principle, is, I take it, a wholly language-dependent possibility. And language is essentially social. Thus the condition for genuinely free individual action is the embedding of the individual in society. Thus, finally, the causal capacities most characteristically and uniquely human are capacities that derive not solely from the internal structure of humans, or human brains, but that depend essentially on the relationship between an individual and society.

None of this should seem surprising to those who take seriously the fundamental biological fact that *Homo sapiens* is a social animal. We should not be surprised that the kind of freedom we possess derives from our being the kind of creatures we are. It may, however, be an unwelcome suggestion for the tradition that connects human freedom with the profoundly individualistic social philosophy and metaphysics dominant in contemporary English-speaking culture. However, I do not mean to imply that autonomy is wholly a social product of which individuals are merely the passive vehicles. Principles or rules do not, as Wittgenstein also famously argued, determine their own application, and rules may be applied with creativity and imagination. And imaginative application or extension

---

[27] Accounts of free will that attach particular importance to a moral dimension in action have also been developed more recently. See, for instance, Watson, 1975; Taylor, 1976.

of rules may increase the range of possibilities open to the members of a society for whom the rule or principle is available. Thus the social construction of language, meaning, and possible principles of action makes possible, but does not fully determine, human agency.

To employ a concept unfashionable in contemporary anglophone philosophy, the relation between individual and society is a dialectical one.[28] The situation of individuals in society in which principles and goals of action, and systems of belief, are articulated is what endows those individuals with the capacities to embrace principles and pursue goals beyond the momentary satisfaction of desire. But the exercise of these capacities in action, argument, or the pursuit of individual goals, in turn can affect society. And such action can expand (or perhaps sometimes contract) the range of capacities and options available to individuals. The debates in which almost all of us play some small part about what principles should govern our behaviour, how our children should be educated, and so on, and the actions we take in support of our views on such matters, not only involve the exercise of individual freedom, but ultimately contribute to affecting and expanding the possibilities for human action. It is this dialectical interaction between the individual and society that grounds individual autonomy. And finally, the pluralistic metaphysics I have defended here and elsewhere shows that there is no philosophical difficulty in taking seriously such a relationship of mutual determination between entities at different levels of structural complexity. Once we do take such a possibility seriously, it should be no surprise that some of the most interesting and puzzling aspects of reality should depend on such interactions rather than merely be the properties of disconnected individual things.

7. Conclusion

The central positive theme that this book promotes is pluralism, and more specifically pluralism in our approach to our own kind. Though a good deal of lip-service is currently paid to pluralism, the

---

[28] I use the term 'dialectical' reluctantly, because I do not want to become involved with the great weight of historical baggage the term has collected. However, I know of no other concept that so accurately captures the relationship of mutual interaction and dependence that I wish to convey.

commitment seldom goes very deep. My own project is to insist that pluralism goes all the way down to the basic metaphysical issues of causality and of what kinds of things there are. This metaphysical perspective makes the kind of narrowly focused scientific projects I have been examining look as philosophically misguided as they have proved empirically unrewarding.

The fundamental error with the programmes that I have criticized in this book is the belief, explicit or implicit, that there is some fundamental perspective that will enable us to understand why people do what they do. It hardly needs insisting upon that it is important that humans evolved and have common ancestors with the other creatures we find around us. And nothing could be more important to us than the organization of society and of the labour of individuals in society in such a way as to provide us with a good deal of what we need and want; no doubt economics has something to tell us about such questions. These are important fragments of the picture that we, the uniquely self-reflective animals, have spent the last few millennia trying to put together. But they are fragments, and trying to make one or even a few such fragments stand for the whole presents us with a deformed image of ourselves. One of the most traditional objections to such one-sided, reductive pictures of ourselves is that they leave no room for human autonomy or freedom. In the present chapter I have tried to show that the philosophical context in which I criticize these reductive views does indeed provide an endorsement of this traditional objection.

I maintain, then, that an adequate view of ourselves, were we to acquire one, would include many parts. It would at least include an account of us as biological organisms with immensely complex functioning parts, and an account of how this functioning gave rise to some of the enormously complex capacities we often exhibit. It would require an account of how our societies function, as general as possible but no more general than the empirical facts permit, and an account of how aspects of social organization contribute to the endowment of human individuals with complex capacities that would be in principle beyond the reach of an isolated member of our species. It will include detailed accounts of some of the most important aspects of our social organization, such as economics, and of the history of our societies and our species. And at a philosophical level it will include an account of the nature and limits of our powers to act autonomously to create real change in the world. Or so I suppose.

But it may well be asked whether this open-ended intellectual shopping list really offers any illumination as to what such a portmanteau account of human life would look like. And it is, I suppose, inevitable that I have no very satisfactory answer to this question. To a considerable extent this book represents an exercise in philosophy as Lockean underlabourer, clearing the ground of rubbish in advance of the construction of a sound edifice of knowledge. I do not consider this task of negligible importance: there is much rubbish to be cleared away. Any bookshop will display a fair selection of sometimes best-selling volumes devoted to simplistic accounts of the essence of human nature, and the authors of these works are constantly to be read or heard peddling their wares in the middle-brow print and broadcast media. One might mention, for instance, Steven Pinker's physically if not intellectually weighty, but very widely noticed, *How the Mind Works.*[29] The view presented therein, that the mind is a computer programmed by natural selection in the Stone Age, is as reductive and simplistic an approach to its topic as anyone is likely seriously to propose, and is as lacking in serious insight into the human condition as such an attempt is likely to prove. Yet this is the work of a respected scientist and is treated with considerable public interest. This points to an intellectual pathology well worth critical attention. But it remains easier to say what does not work than what does.

There is, however, a more positive thesis implicit in my negative arguments. I am suggesting that nothing will serve to provide us with insight into human nature of quite the kind we are currently inclined to imagine. We are tempted, to put it crudely, to think that such insight into human nature could be provided by a text grounded in one or a few comprehensive theories (as the example of Pinker's book illustrates). My most radical opposition to this view is to suggest that we might better think of the cultivation of a skill, the ability to understand, or have insight into, human nature and human life, than the writing of a text. Such a change might reflect the move from the very simple phenomena (relatively, of course: I do not suggest that they were so simple as to be easily understood) which science has so successfully explained, to the vastly more complex phenomena of which human nature is no doubt the most complex of all. The search for simplistic theories such as those of contemporary socio-

---

[29] I discuss this work in more detail in Dupré, 1999a.

biology reflects above all the failure to see this divide. Complexity in this context is not just a matter of very difficult sums that we do not yet know how to solve, but the concurrence of different kinds of factors, each of which may well be complex in this same sense, that we do not know how to fit together. Moreover, there is no reason to suppose it is even possible to fit them together in the systematic, even algorithmic, way that is sometimes assumed. We may be able to construct sets of very simplified problems that we can solve quite effectively, but it is quite erroneous to infer from this that we have discovered a method that will in principle solve any arbitrary problem we might be interested in. This is where I want to discern the limits of science referred to in the title of this book. Without in any way refusing the extraordinary range of knowledge that science has provided for us, there are subject matters that require a more synoptic and integrative vision than the analytic methods of science allow. And here, perhaps, there is the possibility for philosophy to graduate from underlabourer to Queen of the Sciences.

Though I advertise no key to the understanding of human nature, and even doubt whether any such key exists, there are important consequences following from the recognition of the complexity and multi-dimensionality of the problem. We might, for instance, try to promote a more cautious and sceptical attitude towards claims to offer scientific solutions to social and even medical problems. In the first chapter of this book I mentioned pharmacological responses to such psychological problems as Attention Deficit Disorder. I do not suggest (nor consider myself qualified to suggest) that such responses are always or generally inappropriate. I do worry, though, that such solutions will tend to encourage the assumption that such problems are unitary conditions with unitary, generally physiological, causes. This seems to me unlikely. Such a perspective may even encourage the inference from a statistically positive effect of a treatment, to the generally beneficial effect of that treatment. Recognition of the complexity of human behaviour, by contrast, should lead to the expectation that a pharmacological intervention will have effects that will be positive in some cases and negative in others. Refusal to reduce the patient to a physiological problem will reinforce the necessity of attending to the complex particularity of the individual case.

Medicine, indeed, though constantly provided with additional resources by science, remains an art. It must surely remain an art because its object, human health, is both complex and normative. Full recognition of this necessity would surely lead to some significant reevaluation of the goals and methods of medical education. And I think that pluralism has profound consequences for education more generally. It is a cliché that we live in an age of exponentially increasing information, and it is often assumed that this necessitates ever-increasing specialization as the task of mastering more than an infinitesimal fraction of this information becomes more and more daunting. But a proper distinction between information and knowledge or even wisdom might lead to quite the opposite conclusion. If a subject matter can only be understood from simultaneous attention to a variety of perspectives, then knowledge of a subject matter will require access to diverse bodies of information. And perhaps part of what amounts to wisdom is the ability to know what kinds of information or knowledge are needed in application to a particular case.

Happily, the explosion of information has coincided with the growth of information technology. And surely much of the benefit of information technology is that we don't need to accumulate vast quantities of information in our brains, but need only learn to gain access to the information we want as it is stored outside our bodies. The combination of these two developments—the realization, first, that the most important problems we face have many aspects and no simple solutions, and second, that the accumulation of large quantities of information inside the human mind has become largely redundant—should make possible a quite radical reconception of education, especially, perhaps, tertiary education. The correct balance between breadth and depth of education is a very difficult matter. Notoriously, science education is thought to require ever-increasing depth and specificity of focus. Recognizing the limits to scientific methodologies should encourage us to shift the balance towards breadth and, perhaps more important, the skills necessary to integrate insights from a variety of perspectives. Medical education again provides a paradigm. As technical information about diagnosis, therapy, and prognosis becomes increasingly accessible from external sources, we may hope that trainee physicians will be able to devote more energy to the very difficult task of learning to appreciate and promote the total well-being of the patient.

The suggestions of the last paragraphs have, of course, been speculative. I indulge these speculations because I do want to emphasize that the thesis presented in this book is a radical one. The triumphant achievements of science in the last few centuries have been extraordinary, and it is hardly surprising that they have to some extent distorted our conception of knowledge as a whole. It is time, nonetheless, to take a more balanced look at what we can expect from science, and at what role may remain for very different approaches to the acquisition of knowledge. Part of the work of achieving this reevaluation is the recognition, which it was the aim of my earlier book (1993a) to accomplish, that the idea of a uniform scientific project gradually spreading its light across the full range of our interests, is a myth. The present book aims to reinforce this message by considering in some detail a domain of enquiry, human nature, for which this myth is particularly inappropriate and unfortunate. Much of the importance of this project is the negative one of immunizing ourselves against the worst excesses of scientism. But in these concluding speculative thoughts I want to suggest that there are more exciting intellectual vistas to explore beyond scientism.

Bibliography

Allen, E., et al. 1975. 'Against "Sociobiology".' *New York Review of Books*, 13 November, pp. 182, 184–6. Reprinted in Caplan (1978).

—— 1976. Sociobiology Study Group of Science for the People. 'Sociobiology—Another Biological Determinism.' *BioScience* 26, no. 3.

Anderson, E. 1993. *Value in Ethics and Economics*. Cambridge, Mass.: Harvard University Press.

Anscombe, G. E. M. 1971. 'Causality and Determination.' An inaugural lecture delivered at Cambridge University. Cambridge: Cambridge University Press.

Barkow, J., Cosmides, L., and Tooby, J., eds. 1992. *The Adapted Mind*. New York: Oxford University Press.

Becker, G. S. 1981; enlarged edition 1991. *A Treatise on the Family*. Cambridge, Mass.: Harvard University Press.

Blackmore, S. 2000. *The Meme Machine*. Oxford: Oxford University Press.

Block, N. 1995. 'How Heritability Misleads About Race.' *Cognition* 56: 90–128.

Boyd, R., and Richerson, P. J. 1985. *Culture and the Evolutionary Process*. Chicago: University of Chicago Press.

Brandon, R. N. 1990. *Adaptation and Environment*. Princeton, N.J.: Princeton University Press.

Broad, C. D. 1952. *Ethics and the History of Philosophy*. London: Routledge & Kegan Paul.

Buller, D. J., ed. 1999. *Function, Selection, and Design*. Albany, N.Y.: SUNY Press.

Buss, D. 1994. *The Evolution of Desire*. New York: Basic Books.

Butler, J. 1990. *Gender Trouble: Feminism and the Subversion of Identity*. London: Routledge.

Caplan, A. L., ed. 1978. *The Sociobiology Debate*. New York: Harper & Row.

Cartwright, Nancy. 1983. *How the Laws of Physics Lie*. Oxford: Oxford University Press.

—— 1989. *Nature's Capacities and their Measurement*. Oxford: Oxford University Press.

—— 1999. *The Dappled World*. Cambridge: Cambridge University Press.

Cavalli-Sforza, L., and Feldman, M. 1981. *Cultural Transmission and Evolution: A Quantitative Approach.* Princeton, N.J.: Princeton University Press.

Chalmers, D. 1996. *The Conscious Mind: In Search of a Fundamental Theory.* Oxford: Oxford University Press.

Cheng, P. W., and Holyoak, K. J. 1989. 'On the Natural Selection of Reasoning Theories.' *Cognition* 33: 285–313.

Chisholm, R. M. 1964. 'Human Freedom and the Self.' The Lindley Lecture, Department of Philosophy, University of Kansas. Reprinted in Watson (1982).

Churchland, P. M. 1995. *The Engine of Reason, the Seat of the Soul: A Philosophical Journey into the Brain.* Cambridge, Mass.: MIT Press.

Churchland, P. S. 1986. *Neurophilosophy: Toward a Unified Science of the Mind-Brain.* Cambridge, Mass.: MIT Press.

Collier, J. F., and Rosaldo, M. Z. 1981. 'Politics and Gender in Simple Societies.' In Ortner and Whitehead (1981): 275–329.

Cosmides, L., and Tooby, J. 1987. 'From Evolution to Behavior: Evolutionary Psychology as the Missing Link.' In Dupré (1987a): 277–306.

Daly, M., and Wilson, M. 1988. 'Evolutionary Social Psychology and Family Homicide.' *Science* 242: 519–24.

Davidson, D. 1970. 'Mental Events.' In L. Foster and J. Swanson, eds., *Experience and Theory.* Amherst: University of Massachusetts Press. Reprinted in Davidson (1980).

—— 1980. *Essays on Actions and Events.* Oxford: Oxford University Press.

Dawkins, R. 1976. *The Selfish Gene.* Oxford: Oxford University Press.

—— 1982. *The Extended Phenotype.* Oxford: Oxford University Press.

—— 1986. *The Blind Watchmaker.* New York: W. W. Norton & Co.

Dennett, D. 1995. *Darwin's Dangerous Idea.* New York: Simon & Schuster.

Dupré, J., ed. 1987a. *The Latest on the Best: Essays on Evolution and Optimality.* Cambridge, Mass.: MIT Press/Bradford Books.

—— 1987b. 'Human Kinds.' In Dupré (1987a): 327–48.

—— 1990. 'The Mental Lives of Non-Human Animals.' In M. Bekoff and D. Jamieson, eds., *Interpretation and Explanation in the Study of Behavior: Comparative Perspectives.* Boulder, Colo.: Westview Press, 428–48. Reprinted in M. Bekoff and D. Jamieson, eds., *Readings in Animal Cognition.* Cambridge, Mass.: MIT Press, 1996, 323–36.

—— 1992. 'Blinded by Science: How Not to Think about Social Problems.' *Behavioral and Brain Sciences* 15: 382–3.

—— 1993a. *The Disorder of Things: Metaphysical Foundations of the Disunity of Science.* Cambridge, Mass.: Harvard University Press.

—— 1993b. 'Could There be a Science of Economics?' *Midwest Studies in Philosophy* 18: 363–78.

—— 1994. 'Against Scientific Imperialism.' *PSA 1994* 2: 374–81.

—— 1998. 'Normal People.' *Social Research* 65: 221–48.

—— 1999a. Review of Stephen Pinker, *How the Mind Works*. *Philosophy of Science* 66: 489–93.

—— 1999b. 'On the Impossibility of a Monistic Account of Species.' In R. A. Wilson (1999).

—— 2001a. 'In Defence of Classification.' *Studies in the History and Philosophy of the Biological and Biomedical Sciences* 32: 203–19.

—— 2001b. 'Economics without Mechanism.' In U. Maki, ed., *The Economic World View: Studies in the Ontology of Economics*. Cambridge: Cambridge University Press, 308–32.

—— and Cartwright, N. 1988. 'Probability and Causality: Why Hume and Indeterminism Don't Mix.' *Noûs* 22: 521–36.

—— and Gagnier, R. 1999. 'The Ends of Economics.' In M.Woodmansee and M. Osteen, eds., *The New Economic Criticism*. New York: Routledge, 175–89.

Durham, W. H. 1978. "Toward a Coevolutionary Theory of Human Biology and Culture". In Caplan (1978).

—— 1991. *Coevolution: Genes, Culture, and Human Diversity*. Stanford, Calif.: Stanford University Press.

Eells, E. 1991. *Probabilistic Causality*. Cambridge: Cambridge University Press.

Ellis, B. J. 1992. 'The Evolution of Sexual Attraction: Evaluative Mechanisms in Women.' In Barkow, Cosmides, and Tooby, eds. (1992): 267–88.

Fausto-Sterling, A. 1985. *Myths of Gender*. New York: Basic Books.

Feyerabend, P. 1975. *Against Method*. London: New Left Books.

—— 1978. *Science in a Free Society*. London: New Left Books.

Frank, R. H., Gilovich, T., and Regan, D. T. 1993. 'Does Studying Economics Inhibit Cooperation? *Journal of Economic Perspectives* 7: 159–71.

Frankfurt, H. 1969. 'Alternate Possibilities and Moral Responsibilities.' *Journal of Philosophy* 66: 829–39.

Freeman, D. 1983. *Margaret Mead and Samoa: The Making and Unmaking of an Anthropological Myth*. Cambridge, Mass.: Harvard University Press.

Gagnier, R. 2000. *On the Insatiability of Human Wants: Economics and Aesthetics in Market Society*. Chicago: University of Chicago Press.

Galison, P. 1987. *How Experiments End*. Chicago: University of Chicago Press.

Garfinkel, A. 1981. *Forms of Explanation*. New Haven, Conn.: Yale University Press.

Gauthier, D. 1986. *Morals by Agreement*. Oxford: Oxford University Press.

Gilligan, C. 1982. *In a Different Voice*. Cambridge, Mass.: Harvard University Press.

Godfrey-Smith, P. 2000. 'On the Theoretical Role of "Genetic Coding".' *Philosophy of Science* 67: 26–44.

Gould, S. J. 1977. 'Biological Potentiality vs. Biological Determinism.' In *Ever Since Darwin: Reflections in Natural History*. Harmondsworth, Middlesex: Penguin Books.

—— and Lewontin, R. C. 1979. 'The Spandrels of San Marco and the Panglossian Paradigm: A Critique of the Adaptationist Programme.' *Proceedings of the Royal Society of London* 205: 581–98.

Greenstein, B. 1993. *The Fragile Male*. New York: Birch Lane Press.

Griffiths, P. E., and Gray, R. D. 1994. 'Developmental Systems and Evolutionary Explanations.' *Journal of Philosophy* 91: 277–304.

Gross, P. R., and Levitt, N. 1994. *Higher Superstition: The Academic Left and its Quarrels With Science*. Baltimore: Johns Hopkins University Press.

Hacking, I. 1999. *The Social Construction of What?* Cambridge, Mass.: Harvard University Press.

Hamilton, W. D., and Zuk, M. 1982. 'Heritable True Fitness and Bright Birds: A Role for Parasites?' *Science* 218: 384–7.

Harré, R., and Madden, E. H. 1975. *Causal Powers*. Oxford: Blackwell.

Herrnstein, R. J., and Murray, C. 1994. *The Bell Curve: Intelligence and Class Structure in American Life*. New York: The Free Press.

Hull, D. L. 1974. *The Philosophy of Biological Science*. Englewood Cliffs, N.J.: Prentice-Hall.

—— 1988. *Science as a Process: An Evolutionary Account of the Social and Conceptual Development of Science*. Chicago: University of Chicago Press.

—— 1989. *The Metaphysics of Evolution*. Albany, N.Y.: SUNY Press.

Hume, D. 1739–40. *A Treatise of Human Nature*. 2nd ed. Oxford: Oxford University Press, 1978.

—— 1748. *An Enquiry Concerning Human Understanding*. Reprint, Indianapolis: Hackett, 1977.

—— 1779. *Dialogues Concerning Natural Religion*. Reprint, Indianapolis: Hackett, 1980.

Jaggar, A. 1983. *Feminist Politics and Human Nature*. Totowa, N.J.: Rowman & Allanheld.

James, William. 1884. 'The Dilemma of Determinism.' In *The Will to Believe*. Reprint, New York: Dover, 1956.

Kant, I. 1948. *The Moral Law: Kant's Groundwork of the Metaphysic of Morals*, ed. H. J. Paton. London: Hutchinson.

Kaplan, J. 2000. *The Limits and Lies of Human Genetic Research: Dangers for Social Policy*. New York: Routledge.

—— and Pigliucci, M. 2001. 'Genes "For" Phenotypes: A Modern History View.' *Biology and Philosophy* 16: 189–213.

Kenrick, D. T., and Keefe, R. C. 1992. 'Age Preferences in Mates Reflect Differences in Reproductive Strategies.' *Behavioral and Brain Sciences* 15: 75–133.

Kim, Jaegwon. 1993. *Supervenience and Mind*. Cambridge: Cambridge University Press.

Kitcher, P. 1981. 'Explanatory Unification.' *Philosophy of Science* 48: 507–31.
—— 1985. *Vaulting Ambition: Sociobiology and the Quest for Human Nature.* Cambridge, Mass.: MIT Press.
—— 1993. *The Advancement of Science: Science Without Legend, Objectivity Without Illusion.* New York: Oxford University Press.
Kuhn, T. S. 1970. *The Structure of Scientific Revolutions.* 2nd. ed. Chicago: Chicago University Press.
Lakatos, I. 1978. *The Methodology of Scientific Research Programmes: Philosophical Papers,* vol. 1, ed. J. Worrall and G. Currie. Cambridge: Cambridge University Press.
Lawson, T. 1997. *Economics and Reality.* London: Routledge.
Levitt, N. 1999. *Prometheus Bedevilled: Science and the Contradictions of Contemporary Culture.* New Brunswick, N.J.: Rutgers University Press.
Lewontin, R. C., and Dunn, L. C. 1960. 'The Evolutionary Dynamics of a Polymorphism in the House Mouse.' *Genetics* 45: 705–22.
—— Rose, S., and Kamin, L. J. 1984. *Not in Our Genes.* New York: Pantheon Books.
Lloyd, E. A. 1988. *The Structure and Confirmation of Evolutionary Theory.* Westport, Conn.: Greenwood Press.
—— 1999. 'Evolutionary Psychology: The Burdens of Proof.' *Biology and Philosophy* 14: 211–33.
Mackie, J. L. 1974. *The Cement of the Universe.* Oxford: Oxford University Press.
Magnus, D. 1998. 'Evolution without Change in Gene Frequencies.' *Biology and Philosophy* 13: 255–61.
Maynard Smith, J. 1983. *Evolution and the Theory of Games.* Cambridge: Cambridge University Press.
Mayr, E. 1970. *Populations, Species, and Evolution.* Cambridge, Mass.: Harvard University Press.
Mead, M. 1949. *Male and Female.* New York: Morrow.
Mill, J. S. 1859. *On Liberty.* Reprint, Indianapolis: Hackett, 1978.
—— 1873. *Autobiography.* Reprint, Boston: Houghton Mifflin, 1969.
—— 1875. *System of Logic.* 8th. ed. London: Longmans.
Miller, G. 2000. *The Mating Mind: How Sexual Choice Shaped the Evolution of Human Nature.* London: William Heinemann.
Moore, G. E. 1903. *Principia Ethica.* Cambridge: Cambridge University Press.
Morris, D. 1967. *The Naked Ape.* New York: McGraw-Hill.
National Institutes of Health. 1998. 'Diagnosis and Treatment of Attention Deficit Hyperactivity Disorder.' NIH Consensus Statement 1998, 16–18 November; 16 (2): 1–37.
O'Hear, A. 1997. *Beyond Evolution: Human Nature and the Limits of Evolutionary Explanation.* Oxford: Oxford University Press.

Ortner, S. B., and Whitehead, H. 1981. *Sexual Meanings: The Cultural Construction of Gender and Sexuality.* Cambridge: Cambridge University Press.

Oyama, S. 1985. *The Ontogeny of Information.* New York: Cambridge University Press.

Paley, W. 1802. *Natural Theology: Or, Evidence of the Existence and Attributes of the Deity Collected from the Appearances of Nature.* London: Faulder.

Pérusse, D. 1993. 'Cultural and Reproductive Success in Industrial Societies: Testing the Relationship at the Proximate and Ultimate Levels.' *Behavioral and Brain Sciences* 16: 267–322.

Perry, J. R., Macken, E., Scott, N., and McKinley, J. 1996. 'Disability, Inability, and Cyberspace.' In B. Freeman, ed., *Designing Computers for People: Human Values and the Design of Computer Technology.* Stanford, Calif.: CSLI Publications.

Philipson, T. J., and Posner, R. A. 1993. *Private Choices and Public Health: The AIDS Epidemic in Economic Perspective.* Cambridge, Mass.: Harvard University Press.

Pigliucci, M., and Kaplan, J. 2000. 'The Fall and Rise of Dr. Pangloss: Adaptationism and the Spandrels Paper 20 Years Later.' *Trends in Ecology and Evolution* 15: 66–70.

Pinker, S. 1997. *How the Mind Works.* New York: Norton.

Popper, K. R. 1959. *The Logic of Scientific Discovery.* London: Hutchison.

Rosenberg, A. 1994. *Instrumental Biology or the Disunity of Science.* Chicago: Chicago University Press.

Ruse, M., and Wilson, E. O. 1986. 'Moral Philosophy as Applied Science.' *Philosophy* 61: 173–92.

Scerri, E. 1991. 'Chemistry, Spectroscopy, and the Question of Reduction.' *Journal of Chemical Education* 68: 122–6.

—— 1994. 'Has Chemistry Been at least Approximately Reduced to Quantum Mechanics?' *PSA 1994* 1: 160–70.

Scitovsky, T. 1976. *The Joyless Economy.* Oxford: Oxford University Press.

Searle, J. R. 1995. *The Construction of Social Reality.* London: Penguin Books.

Segerstråle, U. 2000. *Defenders of the Truth: The Battle for Science in the Sociology Debate and Beyond.* Oxford: Oxford University Press.

Sen, A. K. 1979. 'Rational Fools: A Critique of the Behavioural Foundations of Economic Theory.' In F. Hahn and M. Hollis, eds., *Philosophy and Economic Theory.* Oxford: Oxford University Press, 86–109.

—— 1987. *The Standard of Living.* Cambridge: Cambridge University Press.

Short, R. V. 1977. 'Sexual Selection and the Descent of Man.' In J. H. Calaby and C. H. Tyndale-Briscoe, eds., *Reproduction and Evolution.* Canberra: Australian Academy of Sciences.

Singh, D. 1993. 'Adaptive Significance of Waist-to-Hip Ratio and Female Physical Attractiveness.' *Journal of Personality and Social Psychology* 65: 293–307.

Smart, J. J. C. 1978. 'The Content of Physicalism.' *Philosophical Quarterly* 28: 339–41.

Smith, A. 1776. *The Wealth of Nations.* Ed. E. Cannan, New York: The Modern Library, 1994.

Sober, E. 1984. *The Nature of Selection.* Cambridge, Mass.: MIT Press.

—— 1994. *From a Biological Point of View.* Cambridge: Cambridge University Press.

—— and Lewontin, R. C. 1982. 'Artifact, Cause, and Genic Selection.' *Philosophy of Science* 49: 157–80.

—— and Wilson, D. S. 1998. *Unto Others: The Evolution of Altruism.* Cambridge, Mass.: Harvard University Press.

Sokal, R. R., and Sneath, P. H. A. 1963. *Principles of Numerical Taxonomy.* San Francisco: Freeman.

Sterelny, K., and Kitcher, P. 1988. 'The Return of the Gene.' *Journal of Philosophy* 85: 339–61.

Strawson, P. F. 1974. 'Freedom and Resentment.' In *Freedom and Resentment.* London: Methuen. Reprinted in Watson (1982).

Stump. E. 1996. 'Libertarian Freedom and the Principle of Alternative Possibilities.' In J. Jordan and D. Howard-Snyder, eds., *Faith, Freedom and Rationality: Philosophy of Religion Today.* Lanham, Md.: Rowman & Littlefield.

Suppes, Patrick. 1993. 'The Transcendental Character of Determinism.' *Midwest Studies in Philosophy* 18: 242–57.

Symons, D. 1979. *The Evolution of Human Sexuality.* New York: Oxford University Press.

Taylor, C. 1976. 'Responsibility for Self.' In A. O. Rorty, ed., *The Identities of Persons.* Berkeley: University of California Press. Reprinted in Watson (1982).

Thornhill, R., and Thornhill, N. W. 1983. 'Human Rape: An Evolutionary Analysis.' *Ethology and Sociobiology* 4: 137–73.

—— —— 1992. 'The Evolutionary Psychology of Men's Coercive Sexuality.' *Behavioral and Brain Sciences* 15: 363–421.

Tooby, J., and Cosmides, L. 1992. 'The Psychological Foundations of Culture.' In Barkow, Cosmides, and Tooby (1992): 19–136.

—— —— 1994. 'Better than Rational: Evolutionary Psychology and the Invisible Hand.' *American Economic Review* 84: 327–32.

Trivers, R. 1972. 'Parental Investment and Sexual Selection.' In B. Campbell, ed., *Sexual Selection and the Descent of Man.* New York: Aldine de Gruyter, 136–79.

Tucker, R. C., ed. 1978. *The Marx–Engels Reader.* New York: W. W. Norton.

Van Valen, L. 1976. 'Ecological Species, Multispecies, Oaks.' *Taxon* 25: 233–9.

Waring, M. 1988. *If Women Counted.* San Francisco: Harper & Row.

Wason, P. C. 1968. 'Reasoning About a Rule.' *Quarterly Journal of Experimental Psychology* 20: 273–81.

Waters, C. K. 1990. 'Why the Anti-Reductionist Consensus Won't Survive: The Case of Classical Mendelian Genetics.' In *PSA 1990*, ed. A. Fine, M. Forbes, and L. Wessels. East Lansing, Mich.: Philosophy of Science Association.

Watson, G. ed. 1982. *Free Will*. Oxford: Oxford University Press.

—— 1975. 'Free Agency.' *Journal of Philosophy* 72: 205–20. Reprinted in Watson (1982).

Williams, G. C. 1966. *Adaptation and Natural Selection*, Princeton, N.J.: Princeton University Press.

Wilson, E. O. 1975a. *Sociobiology: The New Synthesis*. Cambridge, Mass.: Harvard University Press.

—— 1975b. 'For Sociobiology.' *New York Review of Books*, 11 December. Reprinted in Caplan (1978): 265–8.

—— 1978. *On Human Nature*. Cambridge, Mass.: Harvard University Press.

Wilson, M., and Daly, M. 1992. 'The Man Who Mistook His Wife for a Chattel.' In Barkow, Cosmides, and Tooby (1992): 289–322.

Wilson, R. A. 1999. *Species: New Interdisciplinary Essays*. Cambridge, Mass.: MIT Press/Bradford Books.

Wittgenstein, L. 1953. *Philosophical Investigations*. Oxford: Blackwell.

Woodward, J. 2000. 'Explanation and invariance in the Special Sciences.' *British Journal for the Philosophy of Science* 51: 197–254.

Wright, R. 1994. *The Moral Animal*. New York: Pantheon Books.

Wynne-Edwards, V. C. 1962. *Animal Dispersion in Relation to Social Behaviour*. Edinburgh: Oliver & Boyd.

Index

action, explanation of 32, 117–19
adaptation 77–8
adaptationism 20, 82–3, 146
agent causation 156–7
AIDS, *see* HIV
altruism 125–6, 129
Anderson, E. 129
animals and humans 62–3
Anscombe, E. 10n, 178n
atavism 25–31, 67–8, 93
Attention Deficit Hyperactivity Disorder
    (ADHD) 14–15, 185
autonomy 8, 17–18, 156–8, 161, 174,
    175–6
  moral 177–182
  See also free will

Becker, G. 50n, 127–8, 129, 133–6
Block, N. 30n
Boyd, R. 94n
Blackmore, S. 94n
brains:
  as causes of behaviour 31–8
  development of 29
  evolution of 21, 73, 83
  genetic basis of 21, 29–30
  as information-processing devices 83n
  and minds 33–5
Brandon, R. 37n
Broad, C. 156n
Buss, D. 48–9, 50–4, 56, 63, 65, 66, 67, 87
Butler, J. 44n
Byrne, P. 43n

Cannan, E. 143
capacities 173; *see also* human capacities
Cartwright, N. 12, 14, 136n, 157n, 162n,
    166, 178n
causal completeness 157, 158, 161,
    169–70
  arguments against 162–70
causal powers 162, 170, 173, 176, 178–9

Cavalli-Sforza, L. 94n
ceteris paribus conditions 10, 96, 136,
    162n, 166, 170
Chalmers, D. 155n, 162n
chaos theory 155
Cheng, P. 61
Chisholm, R. 156–7
Chomsky, N. 87n
Churchland, P. M. 6
Churchland, P. S. 6
Collier, J. 81, 107
compatibilism 156, 158, 161, 178–9
contingency 76–7
creationism 75
Cosmides, L. 3n, 31, 38, 39–42, 59n,
    60–1, 71–3, 119n
cultural evolution 94–5, 98–102, 107–9
cultural species 99–102, 107–9
  barriers between 107
culture 37, 38–42
  biological substrate of 109
  diversity of 44–5, 58–9, 73, 93–112
  forces of 96–8
  human capacity for 101
  value of diversity of 109–12

Daly, M. 39, 41–2, 49, 59n, 61–2, 65n, 89
Darwin, C. 19, 79
Darwinism 74, 75, 77
Davidson, D. 161n
Dawkins, R. 26, 27, 39, 46, 47–8, 78–9,
    94n
Dennett, D. 20n, 28, 32, 33n, 74, 77–8,
    79, 82
Descartes, R. 5, 32, 33n
design 77–80, 175
determinism 12, 86, 159–77
  and everyday experience 166–70
  and free will 155–9
  and machines 170–3
  methodological 164, 167
  microphysical 159–65

determinism (*cont.*)
　and organisms 173–7
　and scientific investigation 164–6
　See also causal completeness, genetic
　　determinism
development 8, 80, 82
Developmental Systems Theory (DST)
　28–31, 94–5
　and human nature 95
division of labour 120–1
downward causation 160n, 162
dualism 5, 32–3, 71, 176n
Durham, W. 95n

economics 2–4, 17, 83, 120–3, 128
　as antidote to altruism 130
　and biology 3–4, 137–8
　and gender 149–50
　goals of 147–8
　heterodox 120n
　values in 146–52
economism 50n
education 186–7
Eells, E. 178n
eliminativism 6, 160, 161
Ellis, B. 52, 66–7
empiricism 9, 13, 163–4
environment 13, 15, 72
　and capacities 36–7
　and development 30–1
environment of evolutionary adaptation
　21
epistemic virtue 71, 115
　generality, as an 84
evolution 19, 137
　and biochemistry 109
　constraints on 79
　contingency of 76–80
　definitions of 26, 93
　and essentialism 103–4
　and history 99
　and optimality 42–3
　See also cultural evolution
evolutionary psychology 2, 8, 15–16,
　19–92
　adaptability of 64, 66–7
　and animals 62–3
　and atavism 25–31, 67–68
　and autonomy 179
　emergence of 21
　evidence for 54–63
　and genes 21
　and the genetic fallacy 75–6

normative implications of 85–92
　and reductionism 72–3
　of sex and gender 44–69
　vs. sociobiology 22
experiments 11

fact and value 16, 146
　in economics 146–52
Fausto-Sterling, A. 20n, 63n
Feldman, M. 94n
Feyerabend, P. 4n, 116
folk psychology 32
Frank, R. 130
Frankfurt, H. 180n
free trade 149
free will 18, 132, 154–87
　and determinism 154–9
Freeman, D. 45
function 77

Gagnier, R. 121n, 128n, 137n, 145n, 147n
Galison, P. 11n, 165n
game theory 119, 138
Garfinkel, A. 142n
Gauthier, D. 129
genes:
　for behaviour 2, 3, 39–40
　and brains 29–30
　and evil 49
　and evolutionary psychology 21
　and information 29, 30
　and intelligence 30
　and reductionism 72–3
　selection of 26–8, 93
　for tastes 3
genetic determinism 28–31, 38–40
genetic fallacy 75–6
Gilovich, T. 130
globalization 110–12
Godfrey-Smith, P. 29
Gould, S. 20, 38, 40, 64, 82, 83, 85, 146
Gray, R. 28n
Griffiths, P. 28n
Greenstein, B. 48
Gross, P. 113n

Hacking, I. 131
Haddock, A. 87n
Hamilton, W. 126
Harré, R. 178n
Herrnstein, R. 30n
HIV:
　epidemiology of 123–6

testing for 125
Hobbes, T. 139
Holyoak, K. 61
hormones 1
Hull, D. 85n
human capacities 8, 18, 36–38, 132, 158,
        177, 181–2
    diversity of 37–8
    vs. non-human animals' capacities
        36–7
    and occupations 130
human diversity 16, 93–102, 107–12
Hume, D. 19, 88, 158, 166, 178–9

idealization 134
imperialism, see scientific imperialism
indeterminism 157–8, 167–70
    infectiousness of 168–9
    See also determinism
inflation 150–1
    underlying rate of 151n
information 186
instrumentalism 7

Jaggar, A. 44n
James, W. 158n
jealousy 41–2
Jones, S. 114

Kamin, L. 20n
Kant, I. 132, 179–81
Kaplan, J. 2n, 15n, 20n, 40n, 63n, 76n
Keefe, R. 53, 57
Kenrick, D. 53, 57
Kim, J. 160n
Kinsey Reports 64
Kitcher, P. 20, 27n, 47, 52n, 85n, 135n,
        176n

labour, see work
labour market, see market, labour
labour theory of value 139
Lakatos, I. 67, 113
language 18, 33–6
    and autonomy 181
    as constitutive of the mental 33–5
    as social 35–6, 176–7
laws 96
    historical 76
    statistical 169–70
    See also determinism
Lawson, T. 120n
Leibniz, G. 88n

Levins, R. 87
Levitt, N. 113–16
Lewontin, R. 20, 27n, 64, 82, 83, 85, 87
Lloyd, E. 27n, 61
Locke, J. 35

machines 11–14, 165, 170–3; see also
        mechanism
Mackie, J. 166, 167, 168
Madden, E. 178n
Magnus, D. 27
markets 83–4, 120–4, 146, 149
    diversity of 122–3
    labour 138–9, 141–5
    marriage 127, 133
    for pollution licences 131
    for risky sex 123–5
    See also globalization
marriage 59
    market 127, 133
Marx, K. 87, 139–40, 143
materialism 5; see also physicalism, elimi-
        nativism
mathematics 114
    as obfuscation 133–4, 138n
Maynard Smith, J. 119
Mayr, E. 104, 105
Mead, M. 45
meaning 35
mechanism 7, 11, 173, 175; see also
        machines
meiotic drive 27
memes 94n
mental modules 22, 39, 41–2, 43, 74
    evidence for 42
    integration of 68
methodology 128; see also scientific
        method
Mill, J. S. 111, 116, 141, 143, 166, 168
Miller, G. 49, 56n
models 11–14, 134–6
    and reality 13–14, 134
    robustness of 135–6
modules, see mental modules
Moore, G. 88
Murray, C. 30n

naturalistic fallacy 86, 87–9, 91, 92
Newtonian mechanics 10, 134, 165–6,
        170
norms 118; see also rules

O'Hear, A. 34n

optimality 42–3
organisms 173–7
Ortner, S. 59n
Oyama, S. 28n

pair-bonding 64
Paley, W. 173n
Pareto optimality 146–7
Perry, J. 36n
Pérusse, D. 53
Philipson, T. 123–6, 128, 129
physicalism 5–6, 159
physics 5
    completeness of 5, 7, 8–9, 72
Pigliucci, M. 20n, 40n
pigs 149
Pinker, S. 87, 184
Pleistocene, see Stone Age
pluralism 4, 5 , 16, 17–18, 33, 80–1, 115,
        138–46, 162, 182–3
    consequences of 185–7
Popper, K. 85n
Posner, R. 123–6, 128, 129
principles 131–2, 180–2
prisoner's dilemma 152n
promiscuity:
    of apes 46–7
    of men 46, 47, 64
    of voles 1
    of women 64–6
psychological modules, see mental mod-
        ules

quantum mechanics 155–6, 157, 168

race 101–2
rape 25, 53, 55–6, 62–3, 89–91, 97
rational choice theory 117–38, 153
    and autonomy 179
    and reductionism 131
    values in 146, 152–3
reductionism 6, 7, 71–5, 131, 159–62,
        165
    and monocausality 131
    practical consequences of 14–15
    as regulative ideal 9–10
Regan, D. 130
reproductive investment 45–6, 50
reverse engineering 78, 80
Richerson, P. 94n
Rosaldo, M. 81, 107
Rose, S. 20n
Rosenberg, A. 7

rules 60–1, 96–8, 100–1, 181–2
Ruse, M. 86

Scerri, E. 165
science:
    and religion 4
    organization of 84–5
    definition of 113–15
    See also unity of science
Science Wars 16, 113–16
scientific imperialism 16–17, 74, 82–5,
        119–20, 123–32, 137–8
scientific method 115–16, 136
scientism 1–2, 16, 37, 113–16, 129–36,
        146
    and free will 154
Scitovsky, T. 111n
Searle, J. 34n, 177n
Segersträle, U. 85n
Sen, A. 129, 131, 148
sex 44–69
    and animal behaviour 45–7
    and deception 46, 49, 51
    as economic transaction 50–4, 65,
        123–5
    and fertility 126–8
    and gender 44–5
    risky 123–5, 128
    and violence 49–50
sexual attractiveness 50–4, 67
    and class 53–4
    and fidelity 53
    of men 50–2
    of women 52–4
Short, R. 47n
Singh, D. 52
singles bars 124
Smart, J. 5
Smith, A. 50, 120–1, 130, 139, 140,
        142–5, 149
Sneath, P. 104n
Sober, E. 27n, 38n
sociobiology 19–21, 85–6; see also evolu-
        tionary psychology
Sociobiology Study Group of Science for
        the People 86, 87
Sokal, R. 104n
spandrels 20
species 102–6
    asexual 105
    biological s. concept 104–5
    biological vs. cultural 107–8
    essence of 102–4

phylogenetic s. concepts 105–6
    See also cultural species
sperm competition 47n, 65
standard of living 148
Standard Social Sciences Model (SSSM)
    30–1, 38, 45, 58, 73
Sterelny, K. 27n
Stewart, Justice Potter 113
Stone Age 21, 25–6, 31, 42–3, 61, 81
    stable conditions in 38n
Strawson, P. 178n
Stump, E. 180n
supervenience 160, 161–2
Suppes, Patrick 166n
Symons, D. 47

tastes 3, 179
    for altruism 129
    for children 127
taxation 134–6
taxonomy, *see* species
Taylor, C. 181n
Thornhill, N. 53, 56, 87, 89
Thornhill, R. 53, 56, 87, 89
Tooby, J. 3n, 31, 38, 39–42, 59n, 60, 61,
    71–3, 119n
tragedy of the commons 152n
Trivers, R. 45n, 50

Unity of Science 70–5, 84; *see also* reduc-
    tionism
utility 128–9, 137
    and principles 131

and standard of living 148
of work 143

values:
    social constitution of 152–3
    See also fact and value
Van Valen, L. 105
violence 49–50, 76, 81
    by step-parents 61–2

waist-to-hip ratio 52–3, 55
Waring, M. 149–50
Wason, P. 60
Watson, G. 181n
Whitehead, H. 59n
Williams, G. 26
Wilson, D. 27n
Wilson, E. 19–20, 22, 39, 47, 85–6
Wilson, M. 39, 41–2, 49, 59n, 61–2, 65n,
    89
Wittgenstein, L. 35–6, 177n, 181
Woodward, J. 135n
work 17, 138–46, 150
    as a commodity 141–6
    as fulfilment 140–1
    as the source of value 139–40
    as toil and trouble 139
World Trade Organization 110, 121,
    149
Wright, R. 49
Wynne-Edwards, V. 27n

Zuk, M. 126

Printed in the United States
by Bookmasters

Printed in the United States
By Bookmasters